EVERYONE'S GONE TO THE MOON

JULY 1969, LIFE ON EARTH, AND THE EPIC VOYAGE OF APOLLO 11

Joe Cuhaj

Prometheus Books

Essex, Connecticut

Ⓟ Prometheus Books

An imprint of Globe Pequot, the trade division of
The Rowman & Littlefield Publishing Group, Inc.
4501 Forbes Blvd., Ste. 200
Lanham, MD 20706
www.rowman.com

Distributed by NATIONAL BOOK NETWORK

British Library Cataloguing in Publication Information Available

Library of Congress Cataloging-in-Publication Data

Names: Cuhaj, Joe, author.
Title: Everyone's gone to the moon : July 1969, life on Earth, and the epic voyage of Apollo 11 / Joe Cuhaj.
Description: Lanham, MD : Prometheus, an imprint of Globe Pequot, the trade division of The Rowman & Littlefield Publishing Group, Inc., [2023] | Includes bibliographical references and index. | Summary: "Everyone's Gone to the Moon is a week-by-week journey through July 1969, one of the most pivotal months in human history—in space and here on Earth. This unique book follows the crew of Apollo 11 and NASA as they prepare for the historic first lunar landing alongside the major global events buried beneath headlines covering the historic space mission"— Provided by publisher.
Identifiers: LCCN 2023003936 (print) | LCCN 2023003937 (ebook) | ISBN 9781633888814 (cloth) | ISBN 9781633888821 (epub)
Subjects: LCSH: Project Apollo (U.S.)—History. | Space flight to the moon—History. | Nineteen sixty-nine, A.D. | United States—Social life and customs—1945-1970.
Classification: LCC TL789.8.U6A5 C84 2023 (print) | LCC TL789.8.U6A5 (ebook) | DDC 629.450097309/046—dc23/eng/20230131
LC record available at https://lccn.loc.gov/2023003936
LC ebook record available at https://lccn.loc.gov/2023003937

♾ᵀᴹ The paper used in this publication meets the minimum requirements of American National Standard for Information Sciences—Permanence of Paper for Printed Library Materials, ANSI/NISO Z39.48-1992

To my granddaughter Cora, grandson Stevie,
and the entire Artemis generation.
Have big dreams for the future,
but never forget the past.

CONTENTS

CONTENTS

INTRODUCTION

July 21, 1969, saw humankind make an incredible leap in its evolution, one that would link the intricate stone carvings of ancient people that depicted their awe and reverence of the vastness of space, the words and imagery of literary giants like H. G. Wells, and the scientific genius of Konstantin Tsiolkovsky and Robert Goddard to bring our imagination and theories out of science fiction and into science fact.

Much has been written about the legendary flight of Apollo 11 and the first tentative steps into deep space. In all those hundreds of books, countless articles, and hours of documentaries, the authors and filmmakers make one thing perfectly clear—the world virtually stood still while millions watched in awe as the crew of Apollo 11 landed on the moon and Neil Armstrong took that first giant leap for mankind.

In that brief moment, most of the world was united. While singer-songwriter Jonathan King downplays the title of his 1965 hit record, "Everyone's Gone to the Moon," as a "satire on the Dylan school of meaningful lyrics," it did fit the moment. We had all gone to the moon. But of course, that is a figurative statement, not literal. Even as millions of people around the world from the largest American city to the smallest third-world country stood mesmerized in wonder and amazement at the grainy black-and-white images of Neil Armstrong taking that first tentative step on the surface of another world, our own Earth kept spinning, and at times it felt like it was spinning out of control.

Across the globe wars raged on, poverty was running rampant with families trying to survive day to day, people filled the streets protesting any number of social issues, and an entire generation—later to be known as *Baby*

Boomers—were lifting their voices in song with remarkable music, movies, television, and literature.

I was eleven years old at the time and lived in a very Norman Rockwell-esque suburb in northern New Jersey only minutes from New York City. The town's streets were lined with row upon row of houses with neatly manicured lawns. There was a public swimming pool and a lake that offered endless hours of cooling fun in the heat of summer. On one of the town's many ball fields on any given day of the week throughout the summer you would find me and my friends playing pickup games of baseball or launching model rockets.

During this time, I clearly remember hearing conversations between my parents and their friends as they chatted about the news events of the day and the state of the world over an ice-cold can of Schaefer Beer. My dad was born shortly after my grandparents arrived in America from Czechoslovakia. He was a veteran of World War II and very liberal minded—I think.

My parents' friends and neighbors included all races—Black, Asian, Latino. Several of our neighbors were immigrants from Sweden, Italy, Ireland, and Germany, and our home would be filled with the melodic sound of a wide range of languages and accents. But when a clean-cut teenage friend of my sister's showed up one day with long hair, a beard, and ragged, bead-laden attire, I remember my dad saying, "Come the revolution!" Was he alluding to the fear that was sweeping across the older generation, a.k.a. the "Silent Majority" as President Nixon called them, that the teens of the day were running rampant and would tear apart the country? Or was he siding with the young man's rebellious ways? I wasn't quite sure where he landed on the subject.

My family was very much into history, news, and current events. Days would be filled with the sounds coming out of our portable radio from *News Radio 710, WCBS*, out of the city. It was a never-ending barrage of local and world news that filled our kitchen. Then, every evening after work, my mom and dad would sit quietly on the couch in the living room and take in the news of the day as offered up by the "Most Trusted Man in America," Walter Cronkite. So, as you see, it was difficult for me to live in blissful childhood ignorance of the world around me.

It was a time when my sister had enlisted in the US Navy and was stationed only a few hours away from our home at Lakehurst Naval Air Station. She had Navy friends whom she would bring home with her whenever she was off duty for a few days including a few who had just returned from fighting in Vietnam. Their views of the war were diverse, ranging from seeing it as an honor and their patriotic duty to serve the noble cause of stopping Communism in its tracks to openly questioning why we were even there in the first place. The stories of their experiences truly opened my eyes.

I saw firsthand only a few years earlier the rioting that was sweeping the nation due to racial tensions exploding. My family was driving through the city of Newark, New Jersey, on the Garden State Parkway. As we started going under an overpass, my dad shouted for me to "Get down! Get on the floor!"

I did as he said, not understanding what was going on. I crammed myself down onto the floor and looking up through the window, I saw National Guardsmen on the embankment with their weapons drawn lining the streets. Behind them, buildings were smoldering or left as empty charred hulks.

These images didn't frighten me. They fascinated me and made me want to read more and take in the stories of what led to these moments in history.

I was also a space geek. The early days of crewed spaceflight were exciting, and I would purposely leave late for school (or skip school altogether, but you didn't hear that from me) to watch the launch of the two-man Gemini missions and the Apollo missions to the moon. And I loved building and flying model rockets. The Estes Model Rocket company received a considerable amount of the money I earned from returning old soda bottles or mowing the lawn.

We always had newspapers in the house. Every Sunday I would run down to the local newsstand to pick up copies of the *Daily News* or the local *Bergen Record* for my folks. I would read them, too, but not just for the Sunday comics. I was more into the back pages, the little vignettes of everyday life that gave me a deeper appreciation of those around me and what made up the fabric of life.

In that respect, I wasn't your typical eleven-year-old. The world around me was intriguing. I even had a hobby that was different from anything my friends participated in. I loved listening to shortwave radio. I would get up in the wee hours of the morning and bring my dad's portable Radio Shack Realistic Patrolman SW-60 radio into my bedroom, plug in a random length of wire for an antenna, clamp on a pair of humungous headphones, and listen to radio stations from around the world—Israel, the Soviet Union, Australia, even the tiniest of countries like the Vatican and distant islands like the Seychelles and Maldives.

From these late-night soirees of armchair world traveling, I learned about different cultures and their way of life and heard intriguing news stories that American audiences often didn't hear. I also learned what propaganda was all about and its effects on the world.

Fast forward twenty years to 1989. The world had not yet discovered the internet and streaming channels online. It was still an analog world of flipping through TV channels, which I was doing when I happened upon a CBS News documentary that was airing titled *The Moon above, the Earth Below*. It was a documentary celebrating the twentieth anniversary of the first landing on the moon.

Charles Kuralt and Dan Rather hosted the show, which featured clips from the flight of Apollo 11 and vignettes of what life was like on the historic day when we dared travel to another world. The documentary brought that amazing historic day—July 20, 1969—back to life and what life was like in America on the eventful day through archival news footage and snippets of personal stories from random people they met on that day. Many times, the stories were simple conversations between Americans, like a woman and a man eating hamburgers at a barbecue.

"Money," the man begins, his cowboy hat cocked back on his head, "as far as security, income doesn't make a bit of difference."

The woman takes an immediate defensive posture.

"The hel . . ." She pauses before uttering an expletive. "The heck it doesn't! You have to have enough to live on, to pay your bills, to send your children to school."[1]

There was a long and heartfelt appeal from an immigrant on a New York City street.

"I know people who can't reconcile themselves with this country because it's not perfect," he says with a deep, eastern European accent. "Who the hell are they to expect America to be perfect? It was built by discoverers of the Earth. We created a tremendous country here. How can you expect this to be perfect? You should get down on your knees and praise the Lord. The country has accomplished so much."[2]

From the moment I first saw the documentary, I wanted to expand on the concept and delve deeper into that moment in history, but not only on July 20. I wanted to expand the story to include the entire month of July 1969 and get a fuller picture of life on Earth.

Over the years, as I began looking into the news, pop culture, and life experiences that occurred in that historic month, I quickly realized that, in reality, not everyone had gone to the moon. With that realization came the idea for this book.

With all of the excitement of that eleven-year-old boy fifty plus years prior, I put together a proposal for a book that I would call, *Everyone's Gone to the Moon* and immediately sent it to the acquisitions editor at Prometheus Books, Jake Bonar. Jake had worked with me on my first Prometheus book, *Space Oddities: Forgotten Stories of Mankind's Exploration of Space.*

Jake immediately understood my vision—to write a book that would be a snapshot of what planet Earth was like during that historic month of July 1969, weaving together stories of life on Earth—the news, pop culture, and personal stories of the day—with the preparations and eventual landing of Apollo 11 on the moon.

My goal for this book is to relive the staggering news events of the day—the incredible fight for civil rights that was far from over; the Vietnam War that was relentlessly raging; the day Charles Phillip Arthur George was invested as the Prince of Wales only hours after proceedings were halted due to a bomb scare.

The events of July 1969 weren't all doom and gloom. In fact, the month was rife with remarkable music, television shows, movies, and literature. It was a time when the most famous foursome to ever hit the music world—the Beatles—began disintegrating, the fashions of the day were really making a statement, *Midnight Cowboy* became the first X-rated movie to win an Academy Award, David Bowie sang of an astronaut stranded in space, and old New York Giants and Brooklyn Dodgers baseball fans were praying for a miracle in Flushing, New York, with the hapless New York Mets.

And while all of this was going on and as the Apollo program was hitting its stride, there were other space-related news stories happening, as NASA began preparing for our next step in space—a new space station and space shuttle—while at the same time, the Soviet Union's own space program was literally crashing and burning in an ironic twist to the end of the space race and the race for the moon.

But as I said, a snapshot such as this wouldn't be complete without the stories of everyday people, the people who made up the fabric of life here on Earth. They weren't celebrities or astronauts. They were the everyday people who worked tirelessly, dedicating themselves to bringing a better life to their family. Maybe they were workers at one of the countless NASA facilities around the world, the unsung heroes who made the first moon landing a success. Maybe they were the baker, the mailman, the banker, the hospital worker, the teacher, who religiously went to work day after day and kept America moving forward. Maybe they were just families just trying to make ends meet. Or maybe an eleven-year-old kid who was inspired by the events of the day. Theirs is as much a part of this story as the Apollo 11 astronauts.

This is a snapshot of July 1969, a look back at a momentous month in human history.

I

JULY 1–5, 1969

Ignition Sequence Starts

1

NOT JUST ANY INDEPENDENCE DAY

It was a warm but not unpleasant summer day among the towering skyscrapers of Manhattan with the temperature hovering around eighty-five degrees and low humidity. The streets and sidewalks of the Big Apple seemed almost deserted. Car and pedestrian traffic was so slow that Samuel Grun, a gritty New

Buying ice cream from a street vendor on the 4th of July 1969 in New York City. *Photo by Bernard Gotfryd, July 4th, tourists, N.Y.C. Library of Congress, https://www .loc.gov/item/2020737758/*

York City cab driver, parked his cab along a side street and took extended naps during his shift.[1]

The wealthiest of the city dwellers reclined in their chaise lounge chairs atop luxury high-rises to catch up on their tan, while those who were able to, left the city on this Fourth of July Friday for what would be a rare three-day holiday weekend. These travelers found themselves either ensnarled in massive traffic jams or stuck on commuter trains that were experiencing service problems, especially on the Long Island Railroad, which was thwarting efforts to escape the concrete and steel for the frantic summertime throng on the beaches of Long Island, Coney Island, and the Jersey Shore.

On Manhattan's Lower East Side, fire hydrants had been randomly opened by residents to allow the kids on the block to frolic in their gushing glory. A man simply known as Papo climbed down the brownstone steps in front of his home on Avenue B with a wrench in his hand. As soon as the neighboring children saw what he had in his hand, they all dashed inside to put on a bathing suit or unload their pockets of treasures—yo-yos, bubblegum, baseball cards.

Papo placed the open end of the wrench on a large nut atop the hydrant and with a grunt, managed to turn it. Water began to gush from the steel hydrant. Using his fingers, he strained the brown, rusty water for any stones that might rocket out of the pipeline. When the color of the water turned clear, he was satisfied and opened the valve full.

The children danced and shouted with delight in the gushing water. Papo wanted to have some fun himself. In his other hand, he carried an old, empty beer can that had the top and bottom removed.

Papo watched as a young woman driving a convertible with the top down slowly approached the cascading water. In the car, which bore the unmistakable tan color of a New Jersey license plate, a young man sat in the passenger seat.

She stopped the car and asked Papo if he would let her pass through without getting the interior of the car wet. He assured the couple that they were OK to pass without getting soaked.

A burly truck driver drove up behind the convertible and seeing what was happening, put the truck in park, hopped out of the cab, and stood over Papo. "Don't worry," he said. "I'll make sure he lets you through."

With the truck driver standing ominously next to Papo, the young woman began to slowly drive the car forward, and as she did, Papo wedged the can into the stream of water gushing from the hydrant which turned the can into a water cannon. With the can, he was able to direct the water's flow in any direction he chose.

Carefully, he pointed the chute of water toward the car. It arced gracefully into the air and landed dead center inside the car, soaking its passengers and the interior.

As the car sped off, Papo looked at the truck driver and said, "Some protector you are." The truck driver replied, "Eh, they was from New Jersey."[2]

Across town, four-year-old Anna Arias and her mother walked into Columbia Presbyterian Hospital. They were scared. The girl had what appeared to be a simple viral infection, but just after arriving, her condition took a turn for the worse.

Anna was immediately brought into an exam room as nurse Toby McShay jumped from her station at the check-in counter to rush to the girl's aid. Anna was having seizures. The doctors administered valium, but the seizures continued. They introduced even more medicine intravenously but to no avail. Anna was moved to the cardiac arrest room. The fear was that overmedicating the child could cause her to stop breathing. Twelve doctors, specialists, and McShay worked to try to bring the girl back. One doctor began shouting Anna's name, hoping to invoke a response, and it came. The convulsions stopped and Anna began to cry. She would recover.[3]

Only a few miles north of the city in the town of Litchfield, Connecticut, the small town was not only celebrating the country's 193rd birthday, but also the town's 250th anniversary. In attendance was the mayor and counselor of the town's sister city—Litchfield, England—Mrs. A. G. Millard.[4] Dressed in her official red robe, Mrs. Millard accepted a silver medallion from the city as muskets fired shots in her honor.

On the Mississippi Gulf Coast in the town of Gulfport, the front page of the *Mobile Register* shared a tip from Mr. and Mrs. Chauncey Hinman and their twin girls on how to beat the ninety-degree, humidity-laden heat that had lowered itself upon the coast—sit on a gigantic block of ice. The accompanying photo showed the chilly pair of twins dressed in shorts sitting back to back atop a giant ice cube.[5]

Residents in New Jersey had an unexpected surprise as they prepared to celebrate the nation's birthday. The state legislature overwhelmingly voted to turn four holidays—Washington's Birthday, Memorial Day, Columbus Day, and Veterans Day—into Monday holidays. The governor let the bill sit on his desk without signing it so eventually the deadline passed and it automatically went into law. It was the first time since 1938 that a piece of legislation in the state became law through the failure of the governor to act on it.[6]

In Sandusky, Ohio, it was a slightly warmer than usual Fourth of July as temperatures topped out at eighty-six degrees. A brisk seventeen-mile-per-hour breeze from Lake Erie kept the gathering crowds cool as they waited for the music, games, and fireworks of the holiday to begin.

The sounds of children playing, bands warming up, and the occasional crack of a firecracker filled the air. The stage was set for a perfect Independence Day celebration, that is until 7:00 p.m., when the residents of the Ohio town saw their world literally turn upside down as the weather took a drastic and terrifying turn for the worse.

Then fourteen-year-old Dale Kirkwoods recalled that the sky suddenly and without warning turned a "weird green."[7] A Sandusky teenager, Gary Rice, and his band were one of several that would be playing at the city's annual celebration. He clearly remembers what happened next.

"The soft July breeze stopped being a breeze—it stopped altogether. Then it started [again]. First a roar began that sounded like a horrible freight train coming. From my vantage point, it looked as if a solid gigantic gray waterfall was rapidly approaching from the northwest."[8]

That wind, Kirkwoods said, "picked us up. I couldn't believe this was happening."

Little did the residents of Sandusky know that just a few hours earlier, a tremendous storm had begun brewing over southern Michigan, and when it moved out over the western waters of Lake Erie, it literally exploded in strength and turned into a rare but deadly line of storms that became known as the Ohio Fireworks Derecho.

Derecho is a Spanish word that means "straight" as in "straight-line winds." The word was first used by University of Iowa physics professor, Dr. Gustavus Hinrichs in 1878, to describe this unique weather phenomenon.

In layman's terms, a derecho occurs when the wet air of a thunderstorm meets the drier air that surrounds it. The water in the thunderstorm quickly evaporates, cooling the air around it. The cooler air is denser and rapidly sinks to the ground creating extremely strong winds or downbursts. In a derecho, there could be dozens of these downbursts occurring at the same time each with winds averaging 58 to 110 miles per hour and the line of storms stretching over 240 miles.

The Ohio Fireworks Derecho slammed into the shoreline of Sandusky between 7:30 and 8:30 p.m. with winds between 80 and 110 miles per hour. Torrential rain, hurricane-force winds, and tornadoes battered the town for hours. By the time the storm had passed, eight people were dead, ten thousand homes were damaged or destroyed, seven hundred boats had swamped in Lake Erie, and one hundred people had to be rescued by the Coast Guard.

But the storm wasn't finished. It continued its swath of destruction southeastward into the town of Perry, near Cleveland, where it spawned several tornadoes that killed one and injured eighty others.

As the line of storms continued its march, it created rainfall totals that were off-the-charts. Wayne County saw nine inches of rain in less than fifteen hours, while the town of Wooster recorded fifteen inches. Fifty miles southwest of Akron, Killbuck Creek crested more than twenty feet above normal causing over two dozen deaths.

When the storms finally subsided over Pennsylvania the following day, forty-one people were dead with an additional five hundred injured. The total cost of the storm was estimated to be $65 million, or $1 billion in today's money. It remains one of the costliest and deadliest storms in Ohio history.

In Key Biscayne, Florida, residents lined the streets of the tiny barrier island off the coast of Miami for their annual Fourth of July parade. Leading the procession this year was a very special guest—the president of the United States.

Only six months into his term in office, President Richard Nixon and his wife, Pat, served as the grand marshals for the event. Arriving at the viewing stand reserved for dignitaries, the Nixons stood and waved at the rows of floats, marching bands, and baton twirlers passing by. Occasionally the president jumped into the crowd to shake hands with spectators and talk with children, alarming his secret service detachment.

High above, twelve Air Force Phantom Jets performed a flyby while the president ducked and dodged pieces of hard candy that parade participants were throwing to the crowd. Meanwhile, forty members of the Marine Band, who had to be specially flown in for the event, had to leave almost as soon as they arrived so they could fly back to Washington for that city's annual celebration.

Nixon commented that it was "good to be reminded that, while some Americans were dissenters, the majority had faith."[9]

Indeed, there were dissenters across the country on this Independence Day who felt disenfranchised and alienated from all of America's promises as the nation prepared to take the boldest step mankind had ever taken. There were two faces to the United States in July 1969. One was the face of rural and suburban America, whose hearts swelled with patriotic pride as they unfurled the Stars and Stripes and hung bunting from porch railings. The other was the face of turmoil, an America trying to navigate through dire social and political issues that tinged this year's celebration with despair and anger.

It was hard to ignore the fact that the United States was at a crossroads. For as many towns celebrating Independence Day, there were many more in a state of mourning that the promises the Founding Fathers had made were not being kept, in particular that all men are created equal.

In Black communities across the land, debates were raging about the meaning of the holiday. When asked about flying the flag in observance of Independence Day, baseball Hall of Famer Jackie Robinson said, "I wouldn't fly the flag on the Fourth of July or any other day. When I see a car with a flag pasted on it, I figure the guy behind the wheel isn't my friend."[10]

In Harlem and in urban centers across the country, the red, green, and black flag of the Flag for All African People hung from windows. An African American speaker at a gathering in New York City told a *New York Times* reporter that the colors had specific meaning: "Red for the blood we have shed, black for our blackness, and green for our land."[11]

It was a country that was struggling with its identity, struggling to live up to its ideals, praying for the future as they rode the rockets of Apollo 11 to the moon. But it wasn't only the United States that was experiencing these pains. Countries around the world had their own growing pains to contend with. As July 1969 began, social, political, and geopolitical norms were being challenged, and the world map was being reshaped.

The hot summer month of July 1969 was going to be a complicated one.

2

JUST LOOK...
JUST WATCH

Changing schools at any time, whether it is to a new town or to a completely different setting in the same town, is often difficult for children. It's like starting over: new friends to be made; new rules and procedures to follow; a new culture to be introduced to.

Much like all school systems across the country, the Syracuse Public School System in New York State was working through the upheaval known as desegregation in the mid- and late 1960s, during which time laws and barriers that relegated ethnic groups to separate public facilities, including schools, were being lifted.

Whether you were Black or white, this sudden mixing of the races often caused racial tension that engendered the opposite of the desired effect. While the policy did equalize education across all races, the children would often feel uncomfortable and alienated in their new surroundings and environment, which made learning even more difficult. One Black student being transferred into a formerly all-white Syracuse school said of the experience, "When we go outside, the white kids don't want to play with us; they call us all kinds of names." Another said how uncomfortable social studies class could be. "The teacher will say something about Negro kids and everybody looks at you."[1]

But it wasn't only the students who felt the tension. Often there was tension between teachers and their new students as well. "We had heard about [the Black students'] rowdiness," one teacher said. "We were wondering would . . . they give us trouble?" Another teacher interviewed for a study on Syracuse desegregation said, "I think this was the most pressing issue for the faculty, wondering what would come out of the Negro children compared to our own."[2]

Desegregation of the Syracuse school system was the farthest thing from young Joe Formichella's mind as he transferred from the familiarity of St. James Parish to a Syracuse public school.

"It was a culture shock," Joe says, "but not in a racist way by any means.[3] It was the school year 1966–67. Having only known the Catholic bubble of our parish, St. James, where it's pretty much all one tribe, everybody knows everybody, all about their families, etc., I land in public school and the first thing I sensed were the groups, or gangs, or cliques."

The "new kid" felt as if he didn't fit in with orangish hair, glasses, and braces. As Joe puts it, he was a "spectacle with any number of targets."

"The thing I most remember about that first year," Joe continues, "was that I had to fight to get home seemingly every day for the first few weeks. [A] big freckle-faced kid took immediate homicidal offense to me and would block my passage once off of school property and I'd have to fight. Never did figure that out, but the fact that it was allowed to happen—by any interested party, myself included—was accepted as almost routine, seems to have fostered an operating principle of 'you're on your own pal.'"

Joe was one of eight children in the Formichella family. His father, James Anthony Formichella, flew reconnaissance gliders during World War II, but like so many veterans, he would never discuss his experiences during the war and kept his feelings and impressions of the conflict to himself.

"He would never talk about the war," Joe said. "Once I took him to the Vietnam War Memorial in Washington, D.C., and finally asked him, 'Why don't you talk about the war.' All he said was, 'It was a different war.' That was the end of the conversation."

Joe's family was devout Catholic, and he paints a vivid picture of what life was like in the 1960s growing up with seven siblings.

"We had a small black-and-white television and a small living room," he said. "With eight kids we would have to fight for a seat so we could see the TV. And there was a specific TV hour when we were allowed to watch it.

"We lived in a small house. The washer and dryer were located in the dining room right beside the chair where my dad always sat for dinner. The dryer was very noisy and with eight kids it would be running all day. We would all sit down for dinner, and when it was time to begin and my father was preparing to do the benediction, he would lean back in his chair and open the dryer door so it would stop running. When dinner was finished, he would lean back and close it again to restart the machine."

Joe's father was a die-hard New York Yankees fan. On occasion, he would load the family into the car and make the five-hour drive from Syracuse to New

York City, where he would randomly pick an obscure place to go. "There's a deli here we haven't been to in a while," or "Let's go to such-and-such a place." It was all an excuse to drive into the Bronx where they would pull up to the original Yankee stadium. The old stadium used to have a gap in the outfield fence where people on the outside could look inside the "House That Ruth Built" and gaze in wonder at the deep green field of dreams where the legends of the game—Ruth, Maris, Mantle, Gehrig—had all played.

"Just look," Joe's father would say. "Just look."

That was a catch phrase Joe's father often used. "Just . . ." fill in the blank. When it came to the country's first journey to the moon, the elder Formichella was all in. He had an undeniable exuberance that was contagious when it came to the first lunar landing. To him, it was a noble and patriotic endeavor, not only because of the technical achievement, but also because the country's first Catholic president, John F. Kennedy, had set the course.

"When they [Apollo 11] landed on the moon, I remember my dad saying, 'Just watch. Just watch'."

As Neil Armstrong and Buzz Aldrin walked on the moon, Joe remembers walking outside into the balmy Syracuse summer evening and looking up at the moon. Somehow, it made him feel closer.

On the other side of the world, Frank White saw the first lunar landing and the strife the United States was going through in the lead up to the landing in a completely different light.

"I was attending Oxford on a Rhodes Scholarship in 1969 doing graduate work," White said. "[The previous year] was a chaotic time in the United States and it was a difficult time for Americans living abroad. A lot was happening, but you couldn't do anything about it."[4]

The author of the book, *The Overview Effect*, White remembers 1968 and 1969 vividly as years that reverberated across the world. "The trust and belief in the country began [to erode] with the assassination of President Kennedy and grew with the escalation of the Vietnam War by President Johnson when he promised that he would not do so.

"Nineteen sixty-eight brought everything to a head. In January, the Tet offensive occurred [in Vietnam] where the National Liberation Front and the North Vietnamese created a nationwide attack [on South Vietnam and US personnel and bases]. It shifted opinion of those who thought the US would win [the war]. Suddenly it looked like maybe not."

Protesting the war became the norm, often turning violent. Then there were the successive assassinations of Dr. Martin Luther King Jr., which set city streets on fire, and Bobby Kennedy. It left the young man asking, "Will this ever end?"

Then something spectacular happened at the end of 1968—the flight of Apollo 8, the first time men flew to the moon. It was a mission that wasn't supposed to happen, but NASA was feeling the pressure that the Russians may beat them to the moon. In one Hail Mary, NASA decided to rearrange the missions and send Apollo 8 not to land on the moon, but orbit it, taking photos of possible landing sites and testing the spacecraft in deep space.

On that mission, the first color photograph of the blue, white, and brown marble we call Earth rising in the blackness of space above the barren and lonely surface of the moon was taken. It changed everything.

"'Earthrise' was very impactful," White continued. The image was seared into the public psyche. The Earth was fragile and needed protecting. "You can have a similar analog experience on Earth: Leaving your home country [as I did] and looking back across the ocean, you see what's truly there. When I arrived in Europe, I was a believer in the war. When I interacted with people in Europe who didn't see the US in the same way as I did, I started to question. Maybe this thinking was wrong. They didn't assume we [the United States] would always do the right thing and that our cause was noble."

Like thousands of young American men, White wanted to serve his country. He was a lieutenant in the Army Reserve, but his mindset had changed. He no longer believed that the war was right and thought that it couldn't be won. He went on to organize an anti-war movement at Oxford before returning to the United States, where he went to work on Eugene McCarthy's presidential campaign. McCarthy was the only candidate who was for ending the war.

Then came the flight of Apollo 11 and Neil Armstrong taking that first step on the moon. To White, there was a fundamental change in the minds of Americans—our country had done something quite significant, not only technologically but also in the nation's perception of itself.

White was still at Oxford as he watched the first moonwalk on a black-and-white television set in the kitchen of a small flat he was renting. With the time difference, he and his friends had to force themselves to stay awake. One thing kept rolling in his mind and millions of Americans as well: if we can do that, why can't we deal with social problems? "It may be a bit simplistic, but it did create that mindset that we can do more."

White summarized his feelings as he watched Armstrong and Aldrin hop across the moon. "It didn't take away the problems [on Earth]. It wasn't a panacea. Apollo was the beginning of a new awareness of ourselves and our place in the universe. As [Apollo] astronauts once said, we are all in this together here on Spaceship Earth. We are all astronauts and need to work together as astronauts do."

3

T MINUS FIFTEEN DAYS

A brilliant golden sunrise peaked over the horizon of the Atlantic Ocean, its resplendent rays showering a monolithic tower that stood stoically on its shoreline. The tower was a sleeping giant, poised on the Florida coast ready to take humankind on its greatest voyage of exploration. In only fifteen days, this behemoth of a machine would come alive. The sound of white billowing vapors being vented from its enormous tanks would make the beast sound as if it were breathing, as if it were alive.

Before long, workers would flee the vehicle leaving it standing alone on the shoreline. Only three men would dare to be near the machine and they were going to ride on top of it. In only fifteen days, the base of the massive monolith would be engulfed in a hellish ball of flame. Then it would slowly break free of the bonds that held it securely to Earth and rise skyward.

But on this date—July 1, 1969—the mammoth Saturn V moon rocket stood silently, bathing in the bright Florida sunshine. Launch pad 39A was abuzz with activity as technicians and engineers readied this stack of complex machinery for its date with destiny. The tiny capsule sitting atop the stack would protect the astronauts on their quarter-of-a-million-mile journey through the inhospitable environment of space. Hidden away, just below that capsule, sat a spidery-looking landing craft that would take two of the astronauts to the moon's surface.

The technicians, engineers, and staff from the National Aeronautics and Space Administration (NASA), the government agency tasked with making this dream a reality, as well as countless contractors from around the country had only fifteen days to ready the vessel for its great adventure and ensure that the crew of Apollo 11—Neil Armstrong, Buzz Aldrin, and Michael Collins—had every possible chance of landing safely on another world and returning safely to Earth.

Confidence was riding high at the Kennedy Space Center that the launch of Apollo 11 would be a success. After all, the Saturn V rocket had already been successfully launched five times—two times unmanned and three times with men onboard (Apollo 8, 9, and 10). But before this particular Saturn V could be launched, there were a few loose ends that needed to be tied up, beginning with the actual objectives for the flight.

President John F. Kennedy had set the goal eight years earlier when he startled NASA and the world by proclaiming that the United States would "land a man on the moon by the end of the decade and return him safely to Earth." That was easier said than done, considering that when the president made the declaration, the United States had only sent one man into space in a tiny Mercury capsule, a short fifteen-minute up-and-down suborbital flight that barely took astronaut Alan Shepard past the recognized threshold of space, an altitude of sixty miles.

While the mission of Apollo 11 itself and the mechanics of getting three men to the moon had all been ironed out, the actual mission goals for this first lunar landing were not as concrete. The official wording of the goals had been bandied about since 1965, but up to this point, the mission statement was only in terms of the entire mammoth Apollo program. On July 1, NASA released its final draft of what the objectives for Apollo 11 would be. Mission planners minced no words. In fact, they only used twelve words to summarize the mission: "Perform a manned lunar landing and return. Perform selenological inspection and sampling."[1]

Besides the mission statement, NASA still had to decide how the astronauts would commemorate the historic event once—and if—Aldrin and Armstrong had safely landed on the moon. A special Committee on Symbolic Activities for the First Lunar Landing was appointed by director Thomas Paine to make the final decision. The committee was instructed to select symbolic activities that would not jeopardize crew safety or interfere with mission objectives; that would "signalize the first lunar landing as an historic forward step of all mankind that has been accomplished by the United States"; and that would not give the impression that the United States was "taking possession of the moon" in violation of the Outer Space Treaty.[2]

After careful consideration, NASA's press office announced that three items would be left on the moon to celebrate the achievement: A commemorative plaque, messages from world leaders, and an American flag.[3]

The plaque would be a simple design but one that would capture the significance of the moment perfectly. Deputy NASA administrator Willis Shapley was asked to chair a committee to work out the wording for the plaque. With suggestions from the Smithsonian Institute, Library of Congress, and the National Archives,[4] Shapley eloquently encapsulated the historic moment using only twenty words to do so: "Here men from Earth first set foot on the moon, July 1969 A.D. We came in peace for all mankind."

Beneath an engraving of the Earth's two hemispheres, the plaque was signed by the three astronauts and President Richard Nixon. It would be attached to the front leg of the lunar lander that would transport Armstrong and Aldrin to the moon's surface. During the flight to the moon, the plaque would be covered with a metallic plate to protect it from damage during its long journey. The plate would then be removed by the two astronauts during a brief ceremony on the lunar surface during which they would read the plaque to the world.

The messages of goodwill from world leaders were solicited by NASA Administrator Thomas Paine. In all, seventy-four messages, including one from President Nixon, were received. Each one proclaimed goodwill and peace between nations. The messages were all handwritten or typed then made into images that were reduced two hundred times smaller than their original size to about the size of a pinhead. The images were then etched onto a silicone disc no larger than the size of a US half dollar using a process similar to etching circuit boards and chips. The messages were so small that they appear only as a small dot on the disc.[5]

When completed, the disc was sealed in an aluminum capsule to protect it from the harshness of space with the words "From Planet Earth—July 1969" emblazoned on it. Around the rim of the capsule were written the words, "Goodwill messages from around the world brought to the moon by the astronauts of Apollo 11."[6]

The flag, on the other hand, posed several problems both logistically and politically. Logistically, the problem was how do you get a flag to stand unfurled in an airless environment? For NASA engineer Jack Kinzler, the answer was simple enough: create a rigid frame that was lightweight enough for the trip to the moon but would be strong enough to proudly display the Stars and Stripes.

His solution was a flagpole made of aluminum that would ride to the moon in two pieces. The bottom section would be driven into the moon's surface by one of the astronauts with a hammer. The top section would fit into the lower section. The top section would have a crossbar that would hold the flag and would swing out to display Old Glory.[7] The end result only weighed ten pounds.

Politically, it wasn't as simple. No one wanted to make it look like the United States was breaking the Outer Space Treaty of 1967 by claiming the moon for itself. There was much discussion about the flag, including having the crew leave behind either the United Nations flag or miniature flags from every country in the world.

Administrator Paine made the final decision—it would be the American flag planted by the astronauts. All of the other items left behind—the plaque and messages from Earth—would demonstrate that the Americans were not interested in a territorial grab.

Meanwhile, the final touches were being put on the launch as a final "wet" dress rehearsal was performed during which the Saturn V rocket would go through all of the steps, including fueling its massive tanks, as if it were launch day. When completed, the fuel would be emptied, and the rocket would be deemed ready for launch on July 16.

All went well with the wet dress rehearsal until a leak in the connection between the Saturn V and the rocket's service tower was discovered. The connection was part of a system that replenishes the liquid hydrogen in the rocket's third stage prior to launch. Liquid hydrogen boils away in the tank (the white gas that vents from the rocket while on the launch pad). The simulated countdown was halted for three hours while engineers researched the problem and discovered that it was simply a loose valve. Once it was tightened down, the simulated countdown continued flawlessly.[8] The test was deemed a success and the rocket was certified for launch.

Out in the Pacific Ocean, the aircraft carrier USS *Hornet* was making final preparations for the recovery of the Apollo 11 astronauts when they splashed down in the warm waters of the Pacific on July 24, while at the same time, the giant, sixty-four-meter deep-space satellite dish located in California's Mojave Desert was deemed ready for a big month. Not only would the antenna be picking up the transmissions and data from the Apollo 11 capsule and lunar lander, but also from the Mariner VI and VII spacecraft that would make the closest flyby of Mars and send back images and data from the red planet at the end of the month.

In the months leading up to the launch of Apollo 11, NASA's assistant administrator for public affairs, Julian Scheer, had been pressuring the manager of the astronaut office, Deke Slayton, to make Armstrong, Collins, and Aldrin more accessible to the media in an effort to make them more human to the public and to build excitement for the upcoming mission. Scheer made several attempts to persuade Slayton to chip away at the astronaut's training time in order to placate the press. One of Scheer's ideas was to have the astronauts and their wives attend a tea for the women of the press corps.

Astronauts Neil Armstrong, Edwin "Buzz" Aldrin, and Michael Collins (left to right), appear on stage enclosed in a booth to protect them from contamination for their final press conference before the launch of Apollo 11. *NASA*

Slayton took a stand, telling Scheer that such expansive ideas would upend the crew's training and, seriously, did anyone really think that at this point they would have to "hard sell" Apollo 11 to the public?

"Homes and wives are personal," Slayton told Scheer, "and landing on the moon does not change that. This is just another mission which may land on the moon first, but definitely will not go anywhere on schedule if we cannot keep the crew working instead of entertaining the press."[9]

Slayton did concede one media opportunity for Scheer. On July 5, the media were granted one last chance to ask the crew the deep and probing questions they thought the world wanted to know during a final press conference.

Neil Armstrong, Buzz Aldrin, and Michael Collins strode out onto a stage at the Manned Spacecraft Center wearing bright blue metal gas masks to prevent them from picking up any viruses that could endanger the crew on the flight to the moon.

The trio stepped inside a three-sided plastic tent that was fifty feet away from the gathered journalists and that had a series of blowers behind it that pushed air from their back out into the auditorium, another precaution to keep germs away.[10]

Sitting at a table within the plastic box, the crew removed their masks and sat quietly and patiently as they fielded questions from the reporters. One reporter from Swedish Broadcasting asked the crew, "You are taking the trip of all trips for mankind. Which place would you like to go to on vacation when you come back to Earth?"

Armstrong, who was a famously quiet and introspective individual and one who always seemed uncomfortable with the press, flashed an uncharacteristically broad smile and turned the question around on the reporter as he referenced his uneasiness at that moment being in front of the press.

"I think that the situation being as it is now," he said sitting straight up, "the place I would like to go most immediately is the Lunar Receiving Laboratory."[11]

The reporters in attendance got the message and let out a hearty laugh.

Besides the repetitive questions about how the astronauts' families were handling the pressures of the upcoming flight and how was it decided who would be the first man on the moon (Armstrong emphatically pointed out that he was never asked who would be the first and that it was a NASA decision), one new bit of news was made during the conference. It was announced that the command module that would remain in orbit around the moon with Collins onboard waiting for Armstrong and Aldrin would be named *Columbia*, a symbol of the United States and the name of the fictional spacecraft that landed on the moon in the Jules Verne book *From the Earth to the Moon*. The lunar module that would land on the moon would be called *Eagle*.

On the other side of the world, the other great space superpower, the Soviet Union, was having mixed success in the space race. First, Japanese diplomats revealed that one of their freighters, the *Dai Shi Chinei*, was struck outside of Soviet territorial waters near Siberia by pieces of a Soviet spacecraft that had deorbited.

According to the ship's captain, two Russian vessels appeared immediately after the spacecraft hit the three-thousand-ton ship to recover the pieces, but the captain and his crew picked up enough of the debris to prove that it really did hit their ship causing serious damage and injuring five crewmembers.

The Japanese government was informed of the incident but chose to keep it a secret so as not to provoke the Soviets, and the issue passed quietly out of sight.[12]

As the United States was making final preparations to meet President Kennedy's goal, the Soviet Union's ambitions to beat the Americans to the moon hit a major stumbling block. Little did the world know that the race to be the first to land a man on the moon had already ended.

Throughout the sixties, the Soviets and Americans were duking it out, each topping the other with ever greater and more daring space feats, both aiming for the ultimate goal—being the first to land a human on the moon. During this, the first week of July 1969, the Russians were preparing to test their answer to NASA's massive Saturn V moon rocket—the N1.

Standing at 344 feet tall, the N1 would weigh over six million pounds at launch. Its first stage alone had thirty engines, and together, all of its five stages could carry over fifty-eight thousand pounds of men and machinery to the moon.

After a mighty explosion destroyed the first N1 rocket in February of that year, the Russians were ready to try again on July 3. All hopes for the Soviet Union to send a man to the moon rested on this test flight.

At 23:18:32 Moscow time, the thirty main engines of the first stage came to life. The massive stack of machinery began to slowly lift off from the launch pad, but in less than ten seconds, the rocket burst into flames with enough force and fire that it destroyed not only the rocket, but the launch complex and surrounding ground facilities as well.

The explosion was hidden from the world, like many Soviet space failures, but the accident put an end to the Soviets' quest to beat the Americans to the moon. When the accident was finally revealed, it was announced that one single bolt in this incredible moon rocket had come loose and was sucked into an oxygen pump causing an automatic shutdown of its engines.

With the Soviet Union out of the race, the Americans were only one flight away from taking home the ultimate prize.

4

WALES HAS HER OWN IDENTITY, HER OWN VOICE

Jutting out from the west coast of England and directly across the Irish Sea from Ireland is the beautiful country of Wales with its rugged mountain peaks and coastline, its vibrant green rolling pastures, and mud flats teeming with birds. And yes, Wales is a country and not a principality of Britain, as its residents will proudly tell you.

It is a country in its own right but is still part of the United Kingdom. Even with their own system of government and laws, in 1969 a segment of the country's population felt that Britain had too much influence over the Welsh people and their lives. From this sentiment arose a new nationalist movement that believed Wales could not truly be free if they had to live by laws set down by a group of British lawmakers in London. Things came to a head on July 1, 1969.

The lineage to be the queen or king of England can be difficult for those across the pond to understand, what with dukes, duchesses, princes, and princesses galore. Aside from the queen herself, there is one title that is quite well known around the world, that of the Prince of Wales. Until recently with the passing of Queen Elizabeth II, the title belonged to Prince Charles, who is now the king of England. But how he received the title of Prince of Wales is a rather winding road.

Upon his mother's ascension as the Sovereign Queen of England, Charles automatically became the Duke of Cornwall. In 1958 when Charles was ten years old, he became the Prince of Wales and Earl of Chester when the queen issued Letters of Patent that declared him such. These letters are official legal documents that are used to bestow titles on members of the royal family. Even though the queen had submitted such a letter, it wouldn't be until Charles turned twenty that the ceremony to officially bestow the title Prince of Wales on him would take place at Caernarfon Castle in Wales. The date was set for the first week of July 1969.

The event should have been uneventful. The public showed no signs of dissatisfaction with the proceedings continuing. In fact, the event coordinator, the Earl of Snowdon, David Armstrong-Jones, told the press that he "couldn't imagine that there were many people against it."[1] Unbeknownst to the earl, several Welsh nationalist groups including the Free Wales Army had vowed to wage a campaign of civil disobedience around the ceremony.

The night before the event, approximately one thousand Welsh nationalist youths had gathered outside the castle and spent the night taunting the future king with songs. When the royal procession made its way into the castle, men were arrested for booing. A youth was threatened with lynching by the estimated quarter of a million spectators gathered outside the castle for throwing an egg at the queen's carriage.[2] But these were minor incidents compared to what the Free Wales Army had planned. Their plan was to set bombs and destroy several strategic properties around Wales in protest of British rule.

Just two days before the investiture ceremony was to take place, three men with the Mudiad Amddiffyn Cymru cell in North Wales—Alwyn Jones, George Taylor, and John Jenkins—planted a series of bombs that would disrupt the flow of water from facilities that pumped fresh water to Britain from Wales as well as along train tracks in the town of Abergele and at a country club near Caernarfvon.

One of the bombs exploded prematurely, killing two of the men, Taylor and Jones, instantly. As the royal entourage proceeded by rail to the castle, the train was halted for fifty minutes as officers inspected a suspicious object next to the tracks. It proved to be nothing, but further down the line, there was an actual bomb lying in wait. A train carrying dignitaries to the investiture, including President Nixon's daughter Patricia, crossed over the bomb's location. The royal train followed soon after, crossing the bomb site as well before stopping three-quarters of a mile past its location where the queen and royal family transferred to waiting vehicles for the final half mile ride to the castle.

The explosion came soon after the royals had exited the train and were on their way to the castle. Spectators waiting at the event site heard the echoing sound but thought it was part of a military gun salute. The train's crewmen said the explosion rocked the cars and initial reports said that a portion of the tracks was demolished, but as it turned out, the bomb went off in a vacant lot near the tracks,[3] and no harm came to the royal family or the invited guests. The investiture went on as planned on July 1.

Two youths were arrested after they were spotted carrying explosives near the rail line. The third member of the trio who initially planned the attacks, John Jenkins, was also arrested. But a question remained: Did these actors actually plan to kill the royal family?

For years after the bombing, the surviving member of the attackers, Jenkins, insisted that they did not intend to hurt anyone. In a 2014 interview with *Wales Online*, Jenkins said, "What we were trying to do was let people see the seriousness of the situation and how Wales would get nowhere while decisions were [being] made in London.

"We never intended to hurt anyone, and I still feel guilty about what happened to the two guys 'at Abergele. I feel that I should have been killed."

Meanwhile, in Africa, the country of Kenya was also experiencing growing pains as it sought its own independence from the United Kingdom. That pain would culminate in the assassination of the man that most people assumed would be the successor to the nation's first president, Jomo Kenyatta—Tom Mboya.

Mboya was born the son of a poor Kenyan farm family in 1930. He was educated at several Catholic missionary schools before he set off for Nairobi, where he enrolled in school for sanitary inspectors. Upon graduation in 1950, he was offered a job as a sanitation inspector. From there, Mboya quickly began his rise in politics and in fifteen short years grew to be one of the most influential activists and politicians in pre-independence Kenya.

In the early 1950s, the country of Kenya was a principality of Britain, and, as in Wales, a Kenyan nationalist movement began to rise. The movement, known as the Kenya African Union, or KAU, but informally as Mau Mau, worked to oppose British rule by using any means at their disposal. The movement consisted primarily of members of the nation's largest tribe, the Kikuyu.

In 1951, the KAU, under the leadership of Jomo Kenyatta, began what became known as the Mau Mau Rebellion during which its members would use guerilla warfare tactics against European landowners across the country to forcibly demonstrate their opposition to Britain. By the end of 1952, the British government had declared a state of emergency, and Kenyatta and five hundred of his rebels were arrested.

Mboya stepped in as the treasurer of Kenyatta's KAU, which effectively placed the young man in the role of the new head of the movement when Kenyatta was arrested. Mboya's first act was to unify and mobilize five of the strongest Kenyan labor unions together to form the Kenya Federation of Labor (KFL), which would demonstrate and protest their strength and determination to split from the Commonwealth.

Mboya went on to attend a year of college at Oxford University, where he studied industrial management, but by the time he returned to his home country in 1960, the rebellion had been effectively dissolved. During that year, over ten thousand of the Mau Mau had been killed, while Kenyatta remained jailed in a Kenyan prison. One of Mboya's first duties upon his return from Oxford was to campaign for Kenyatta's release, which was granted in 1961.

At this time, higher educational opportunities for Kenyan and East African students were very limited. In 1960, Mboya sought to change that situation and embarked on a short tour of North America where he requested scholarships from colleges and universities that would eventually allow hundreds of African students to come to the United States and Canada to earn degrees. One of those who was offered and accepted a scholarship was Barack Obama Sr., the father of the future forty-fourth president of the United States.

Many celebrities and politicians in the United States donated money to help bring the students to their colleges including baseball great Jackie Robinson, singer Harry Belafonte, actor Sidney Portier, and Senator John F. Kennedy, whose family foundation donated $100,000 to the effort.[4]

Two years later in 1963, Kenya gained its independence within the Commonwealth, which simply meant that Queen Elizabeth II was still the head of state over the country. The following year, however, Kenya was declared a sovereign country in its own right with Kenyatta being installed as the new president and Mboya the new Minister for Justice and Constitutional Affairs. Mboya went on to become the Minister for Economic Planning and Development where he laid the groundwork for bolstering the Kenyan economy and stabilizing the footing that would ensure a strong mixed economy in the country's future.

It was more than obvious to many that Kenyatta was prepping Mboya to take his place as president. After all, Mboya was the one person who had continued the liberation movement while Kenyatta was in prison, and he had personally secured Kenyatta's release.

In 1968, Mboya began a campaign to do a little political housecleaning of the Kenyan government. He began accusing several Kikuyu politicians of enriching themselves off the backs of their constituents. Tension began to grow between the politicians and Mboya and eventually, the powder keg exploded. On July 5, 1969, as the thirty-eight-year-old Mboya walked out of a drugstore, a gunman, Nahashon Isaac Njenga Njoroge, shot and killed the political superstar.

The reaction to the assassination was swift. Grieving crowds gathered at the hospital in Nairobi and it quickly turned into a riot. Police had to be brought in to calm the situation.

Nine years prior to his death, one of Kenya's independence leaders, Tom Mboya, is featured on the March 7, 1960, cover of *Time* magazine and is lauded as the brightest rising star in the country. *Author's collection*

The gunman, Njoroge, was later found guilty in a court of law and hanged at Kenya's Kamiti Prison for the crime.

In her book *The Other Barack: The Bold and Reckless Life of President Obama's Father*, Sally Jacobs tells us that Barack Obama Sr. was with Mboya at the drugstore where he was assassinated. According to her account, Obama had joked with Mboya that he had parked the car incorrectly in front of the drugstore and that he might get a ticket. Obama Sr. was the final prosecution witness in Njoroge's trial.

In the Pacific, the country of Malaysia was also embroiled in political turmoil that began in May and became known as the May 13th Incident. But it was much more than a simple "incident."

Malaysia, which up until 1965 included the island of Singapore, was at one time a British colony that gained its independence in 1957. From its early British influence, the country had built a mixed population consisting of 62 percent Malaysians, 21 percent Chinese, and 6 percent Indians. This cultural dynamic led to many conflicts over the years between the Malay and Chinese populations, the latter of which increased exponentially in the 1960s to become an economic and political force to be reckoned with in the country. The majority Malay population feared the rapid growth that the Chinese population was experiencing within the country to the point that in 1965, the Malay government expelled Singapore, leaving Singapore as an independent country. While this action calmed the nation's political concerns for the moment, ethnic tensions between the Malays and Chinese continued to build as the Chinese population gained even more political dominance.

A general election for the Malaysian Parliament was held on May 13, 1969. When the final votes were tallied, the Malaysian Alliance Party had lost many seats to the Chinese Democratic Action Party.[5] The election swiftly moved from the voting booth to violence and deadly clashes erupted on the streets between the two factions.

The fear and violence that swept through the country's towns and cities including the capital, Kuala Lumpur, was described to University of California Berkeley student Kelly Jones by her mother, who was caught in the middle of the conflict.

On May 13th, 1969, my mother and her brothers hid underneath their beds in their Petaling Jaya home while sounds of a race riot raged outside, her parents clutching sticks as their only weapons as they waited by the door for attackers. She was eight at the time. There was a lot to fear that night; throughout the capital city of Kuala Lumpur and its townships, Malays killed Chinese in the streets and in their homes, as well as burned down Chinese shophouses, while Chinese secret societies killed Malays in revenge. When the military arrived, they allegedly fired upon the Chinese as well. Tunku Abdul Rahman, the prime minister of Malaysia, called a state of emergency and locked down all of West Malaysia in curfews, while families who had lost their homes took shelter in a stadium. By the time the riots officially ended, it was a different world in Malaysia.[6]

On May 15, a state of emergency was declared and a media blackout was enforced. The fighting ended the first week of July 1969. When it was over, 6,000 people were left homeless, countless vehicles and buildings were burned, thousands of people were injured, and according to government estimates, 196 people were killed, although many reports put the number much higher at between 1,500 and 2,000.

Over fifty years after the May 13th Incident, tensions between the two factions remain, but overall, peace has come to the peninsula.

Another conflict was also deescalated the first week of July 1969, one that many thought would drag the United States back into another war in Korea. In fact, it was called the Second Korean War.

Thirteen years after the armistice that ended the Korean War was signed by the United States and North Korea in 1953, the North launched a new wave of aggression against the South. In late 1966, North Korean leader Kim Il Sung dispatched hundreds of guerilla soldiers into South Korea. The Demilitarized Zone, or DMZ, which had once been a calm and relatively safe border between the two countries, was virtually wiped out, with over 280 border skirmishes occurring between the two countries in early 1967 and gun battles becoming almost a daily occurrence. Bombings and gunfire hit several US installations including barracks, killing American soldiers.

One of the most daring raids by the North was the attempt to assassinate the president of South Korea, Park Chung Hee. On January 17, 1968, an elite commando group from the North snuck across the border and made their way to the president's mansion, known as the Blue House. A group of loggers spotted the soldiers and alerted officials in the capital city of Seoul.

When the commandos arrived at the Blue House, a South Korean police officer spotted them and confronted the men. A gun battle broke out that resulted in eight South Koreans and three of the commandos dead. The soldiers fled the scene, but a nationwide manhunt tracked the commandos down and killed them. In the process, sixty-eight South Koreans died along with three Americans.[7]

Days later, the North Koreans captured the USS *Pueblo*, which was patrolling in international waters off the coast. The captain and crew were subjected to torture and public humiliation. Demands were being made of the newly elected President Nixon to do something in retaliation, such as dropping a "big juicy bomb" on the North's capital city, Pyongyang. The crew were held captive for eleven months until their release was negotiated. The terms required the United States to admit the ship was in North Korean waters and to apologize for its actions.

In April 1969, the cross-border incidents continued as North Korean MIG fighters shot down a US Navy intelligence plane, killing all thirty-one crewmembers. All sides were preparing for yet another war in Southeast Asia, but just as quickly as it had started in 1966, three years later in July 1969, North Korea began to back down, deescalating its aggression to the point where the state of high alert was rescinded and tensions returned to pre-hostility levels. The Second Korean War had been averted.

5

IF IT HAD BEEN "THE BALLAD OF JOHN, GEORGE, AND YOKO" . . .

If a family wanted to watch their favorite television shows in living color in the sultry summer of July 1969, they would have to shell out some big money. The most affordable console model available was a twenty-inch Admiral set for $429.95.[1] That equates to $3,414 in today's money.

A new television was far from the minds of most Americans as July 1969 began. There were only a smattering of new television shows. Most of the fare over the airwaves was a selection of summer reruns of popular shows that had originally aired in the fall of 1968 and the first few months of 1969. Even the five- to six-hour block of children's cartoon shows on Saturday morning was repeats.

There was one brand-new but short lived cartoon show, however, that made an appearance. It was called *Hot Wheels*, and was the first cartoon based entirely on a line of toys, the wildly popular miniature race cars of the same name that were manufactured by the Mattel toy company.

The cartoon revolved around the hero, Jack "Rabbit" Wheeler, and his Hot Wheels Racing Club. The response to the show by parents was swift and loud. The public was outraged that a children's cartoon show was actually a thirty-minute commercial to sell the toy. It was a not-so-subliminal suggestion for kids to bug their parents to go out and buy Hot Wheels cars.

The outcry was so intense that the FCC passed a regulation banning such marketing practices. The regulation was later overturned in 1983, and the boob tube has been inundated with strategic product placement in cartoon shows ever since.

The world was in an interesting juxtaposition in the summer of 1969, with two generations, the "greatest generation" who had fought through World War

II, running headlong into a new and formidable foe—their own children and the counterculture. While parents and children often disagree on politics and music, there was one unlikely television show that brought the two together— *The Mike Douglas Show.*

Born in Chicago in 1925, Michael Delaney Dowd Jr.'s singing career began when he became a crooner on a cruise ship that sailed the Great Lakes. Following World War II, Douglas signed on as a singer for Kay Kyser's big band singing the big hits "Buttermilk Sky" and "The Old Lamplighter."

In 1961, Douglas was asked to host a new television talk show based out of Chicago. *The Mike Douglas Show* featured local and national celebrities and politicians. In just two years, the show, which had only aired on the one Windy City television station, began to pick up syndication across the country. This necessitated a move so that the show could get bigger celebrities from New York City. In 1965, the show moved to Philadelphia, and that's where the magic truly began.

While Douglas was the host of the show, he invited special guests to be his cohosts. Over the years he had such celebrities and national figures as the Rolling Stones, Gene Simmons of the band Kiss, John Lennon and Yoko Ono, even Richard Nixon. Not only were these celebrities cohosting the show, but they would also book their own guests to appear with them and be interviewed.

This literal bridging of the generation gap was a tremendous success. At the height of its run, the show would have seven million people tune in every day. When asked why he thought the show was such a success, Douglas said it was "because I'm a square."

As the first manned landing on the moon drew near, Douglas's show was still topping the ratings, even though it was only a syndicated show and not on a major network. His cohost that first week of July was the Godfather of Soul, James Brown. Once again, the show featured the opposite ends of the entertainment and political spectrum. Comedians Robert Klein and Al Freeman Jr. appeared, as did the jazz and R&B group the Dee Felice Trio, who had sung backup on several of Brown's records; actress and cabaret singer Julie Budd; and the editor of the African American magazine *Jet,* Robert E. Johnson.

To say the world of music was dramatically changing in the summer of 1969 is an understatement. The youth of America began embracing a new form of concert event—the pop music festival.

Over the Fourth of July weekend, more than one hundred thousand people braved one-hundred-degree heat indexes to rock out at the Atlanta International Pop Festival. Topping the bill were some of the biggest names in rock music at the time—Janis Joplin, Canned Heat, Creedence Clearwater Revival, and Led Zeppelin, to name only a few.

One of the premier jazz festivals was about to make a huge change, and in the eyes of many of its followers, it was a change not for the better. It happened at the granddaddy of them all, the Newport Jazz Festival in Rhode Island.

The story of the Newport Jazz festival begins in 1950 when a twenty-four-year-old college student named George Wein took the money that he had set aside for his education and opened a bar in Boston that would feature live jazz music. His first attempt at opening a club was a failure. The second club struggled, but things changed when he was able to raise enough money to book legendary jazz pianist George Shearing for a performance. With that, word of the club spread like wildfire, and the bar became the place to go in the Boston area for jazz.

In 1953 Wein was approached by Elaine Lorillard, a music lover and a member of the upper-class set who lived in Newport, Rhode Island. Lorillard thought that in the summer, Newport was a boring, dull place, so she asked Wein if he could bring jazz to the island. And he did. The first American Jazz Festival, later to be renamed the Newport Jazz Festival, was held on July 17, 1954. Over two days, the inaugural event held various panel discussions on the art and direction of jazz and brought to the stage many memorable performances by the likes of Eddie Condon, Ella Fitzgerald, and Billie Holiday.

Through the years, the festival slowly integrated new forms of music into the event which brought a broader range of fans in to experience the true American musical art form. But during the festival weekend that began on July 3, 1969, the festival's experiment with fusing jazz with soul and rock music took a drastic turn for the worse.

The weekend would feature the traditional greats of jazz, but new to the lineup were groups such as Sly and the Family Stone, Jethro Tull, James Brown, Led Zeppelin, and Blind Faith. A crowd of twenty-one thousand ticketholders had seats inside the venue while on the outside, a crowd of ten thousand without tickets sat on a hillside with a view of the stage where they could sit and listen to performances.

During Sly and the Family Stone's long set, the hillside crowd went on a rampage. The crowd tore down the wooden fence that surrounded the venue, leaped over railings and chairs, and began battling security guards. Ticket holders had to flee the melee in fear for their lives, among them Rhode Island Governor Frank Licht and Senator Claiborne Pell. Security did manage to keep the crowd from jumping onto the stage, while a contingent of twenty state police were called in to control the situation.[2]

As Sly and the Family Stone left the stage after an hour-long performance, in what was a surreal and visual example of the old adage, "music soothes the savage breast," French jazz violinist Stephane Grappelli and singer Maxine Sullivan took the stage, and the rioters calmed down.[3] Remarkably, no injuries were reported during the fracas.

"The festival was sheer hell," Wein told reporters after canceling the event (Led Zeppelin never performed that weekend.) "The worst four days of my life. . . . The kids destroyed the event and the experiment was a failure."

It was indeed a failure. After receiving the blessing of the Newport City Council the previous year allowing Wein to continue holding the festival, the council changed their referendum, stating that Wein could continue the festival, but rock bands were no longer allowed.

Across the ocean in Britain, the sun was slowly setting on arguably the greatest rock band of all time, the Beatles. On July 4, John Lennon released his first solo single in the United Kingdom, the iconic anthem of the Vietnam War era, "Give Peace a Chance."

Even though the song was completely written by Lennon he gave his longtime friend and bandmate Paul McCartney cowriting credit, which ultimately made the song the last one credited to the songwriting super-duo of Lennon and McCartney. According to Bruce Spizer in his book, *The Beatles on Apple Records*, Lennon made the decision to credit his pal as a way of saying thank you to McCartney for helping him record the Beatles' song, "The Ballad of John and Yoko."

In July 1969, the Beatles were on a downhill spiral with egos and tensions mounting, as evidenced in their 1970 film, *Let It Be*, which was supposed to document how the Beatles wrote and recorded music. Years later, George Harrison recalled the film saying that it was "a great film to show how to break up a band."

Despite all of the disagreements, John and Paul could still work together and produce amazing music. In April 1969, Lennon had penned the rough lyrics to "The Ballad of John and Yoko," the story of the resistance the couple was facing in trying to get married. Once he had the lyrics, he rushed over to McCartney's house to finish it off.[4]

In no time, both the lyrics and music were completed, and the pair sped off to Abbey Road Studios to record it without the other Fabs in attendance. It would just be John and Paul by themselves. When the single was released, George commented on not being included on the recording saying with his dry wit, "If it was [called] 'The Ballad of John, George, and Yoko' I would have been on it."

McCartney later remembered that "John was in an impatient mood so I was happy to help. It's quite a good song. It has always surprised me how with just the two of us on it, it ended up sounding like the Beatles."[5]

Through the years the songwriting credit for "Give Peace a Chance" has caused friction between McCartney and Lennon's widow, Yoko Ono. Sometimes McCartney is credited as cowriter, sometimes he isn't. The music licensing organization, ASCAP, however, will always recognize the song as being written by Lennon and McCartney. In an interview with *Rolling Stone* magazine, a representative with ASCAP said, "We would have to be advised if the split is changed, which would be a good thing to know."[6]

Only days before the release of "Give Peace a Chance," Lennon and his wife, Yoko's daughter Kyoko, and John's son Julian were vacationing in Scotland. Driving in poor weather with low visibility along a narrow Scottish road, Lennon spotted an oncoming car heading straight for the family. Lennon jerked the wheel of his Austin Maxi automobile and careened off the road into a ditch. The family was taken to nearby Golspie Lawson Memorial Hospital where Lennon received seventeen stitches, Yoko fourteen, and Kyoko four. Yoko also injured her back in the incident.

To add insult to injury on this busy day in Beatles' history, the band's song "Get Back," which had been number 1 on the Billboard music charts for five weeks, was knocked out of the number one spot by Henry Mancini and the "Love Theme from *Romeo and Juliet*."

Another iconic British-invasion rock band, the alter-ego of the Beatles, the Rolling Stones, faced turmoil of their own. If the Rolling Stones were known as the "bad boys of rock," then the band's founder, Brian Jones, was the group's original bad boy. In the words of The Who's lead guitar player, Pete Townshend, "[Brian] was on a higher plane of decadence than anyone I would ever meet."[7]

After several drug-related run-ins with the law and his inability to play or even show up for recording sessions due to his drug habit, bandmates Mick Jagger and Keith Richards fired Jones from the band in late June. Three days later, Jones was found dead in his swimming pool.

Rumors and conspiracies about the musician's death immediately began circulating including the belief that it was murder. In the end, the coroner ruled the death of Brian Jones as "death by misadventure."[8]

The band was scheduled to perform a free concert at London's Hyde Park on July 5. With the spirit of "the show must go on," the Rolling Stones took the stage to a crowd of over a quarter million people. Jagger, dressed in a fluffy white outfit that was said to have been borrowed from Sammy Davis Jr., began the show by paying tribute to Jones and started reading from *Adonais*, a poem by Percy Bysshe Shelley:

Peace, peace! He is not dead, he doth not sleep.
He hath awakened from the dream of life.
'Tis we, who lost in stormy visions, keep
With phantoms an unprofitable strife,
And in mad trance, strike with our spirit's knife
Invaluable nothings.

The following day, Jagger was off to Australia to begin filming the movie *Ned Kelly*, the story of an eighteenth-century Australian outlaw. During the filming, Jagger would be wounded in the hand when a prop pistol backfired.

Entertainment in the summer of 1969 was not limited to the music world. It also saw the rise of huge mega-resort casinos in Las Vegas. One in particular would set the stage for those to come when it opened the first week of July 1969—the International Hotel. The hotel was the brainchild of billionaire entrepreneur Kirk Kerkorian and at the time, was the largest such entertainment complex ever built in Sin City.

Born Kerkor Kerkorian in 1917, Kerkorian was the son of Armenian immigrants. His mother was a homemaker, his father a fruit merchant who tried—and failed—at many get-rich-quick schemes that often left the family penniless.

Kirk, as he was later known when he Americanized his name, dropped out of school in the eighth grade, taking up odd jobs and doing a little amateur boxing. During World War II, he became a pilot and flew bomber planes from America to Britain for the Royal Air Force. Following the war, he landed a job flying Hollywood celebrities to Las Vegas, which was quickly becoming a mecca for gambling and entertainment.

Through the years, Kerkorian dabbled in various business ventures and quickly made a fortune in the airline, movie (he once owned MGM Studios), and the gambling megaresort industry. In 1962, he quietly began buying small pieces of property just off Las Vegas Boulevard, better known as the Strip, and in 1968 he made the announcement that he was going to build the largest resort casino ever—the International.

The announcement infuriated the reclusive airplane and movie mogul, Howard Hughes, who had just recently purchased the Sands Hotel, the largest casino in the city at the time. Outraged at being one-upped by this upstart, Hughes announced that the Sands would undergo massive renovations and expansions which, he hoped, would scare off Kerkorian's creditors and end any plans for the International to be built. Hughes's plan failed and Kerkorian was able to gain the credit he needed and the International Hotel was completed.

In an ironic twist of fate, Kerkorian went on to purchase the Sands from the Howard Hughes Estate in 1987 for $167 million, adding to his collection of the International and the MGM Grand which he had built in 1973.[9]

Kerkorian's International Hotel officially opened on July 1, 1969, to huge fanfare, and the following day, singing greats Peggy Lee and Barbra Streisand opened on the hotel's stage for a four-week engagement. At the end of the month, the singers would be replaced by Elvis Presley who was continuing his famous 1968 comeback tour.

The TV airwaves were also ringing out with the sounds and songs of the late 1960s, most notably on a show featuring a group of comic book characters—Archie Andrews, Jughead, Betty, Veronica, Reggie, and Hot Dog—a show that was cleverly titled *Archie*.

The cartoon was so popular that it was renewed for a second season the first week of July 1969. Each *Archie* episode was embedded with a sugary sweet pop song supposedly recorded by Archie's gang of friends who called themselves—once again a "clever" name—the *Archies*. And yes, the songs were literally

Kids in the summer of 1969 could get a free copy of the Archies' song "Sugar, Sugar" on the back of specially marked boxes of Sugar Pops. They were called Flex Discs and would warp almost as soon as they were cut from the box, but the marketing scheme helped push the song to number 1. *Author's collection*

sugary sweet especially their hit song, "Sugar Sugar," which was released this month.

The song was written by Andy Kim and Jeff Barry, recorded by Kim, Toni Wine, Ron Dante, and Ellie Greenwich, and produced by famed record producer Don Kirshner. The song had some ingenious marketing behind it. Kids could get a free 45 RPM record of the song on the back of specially marked boxes of Sugar Pops. The record was actually a very thin vinyl disc that was printed directly on the cereal's cardboard box. Kids would have to cut the disc out of the box in order to play it on a phonograph, much to the chagrin of their parents. But they didn't have long to play it. Inevitably the cardboard would warp and curl up making the disc unplayable. It didn't matter, the gimmick worked, and the song would become a number one hit.

In movie theaters, *The Chairman*, replaced John Wayne and *True Grit* as the top grossing movie, while on newsstands, *Time* magazine led with a cover story featuring Cesar Chavez, the Mexican American who led a boycott on behalf of migrant farm workers demanding bargaining rights for the workers and their right to unionize. *Life* magazine hit the newsstands this week with a special "Off to the Moon" edition featuring Neil Armstrong on the cover.

A FOOTNOTE TO HISTORY

July 1: A half dozen congresspeople were on day five of their six-day "experiment" to find out what it's like to feed yourself and your family on a welfare budget. At this point in time, welfare recipients received what amounted to 18 cents a day to feed their families. The congressmen, who made $42,500 a year, were taking part in a program created by the National Welfare Rights Organization, which was working to convince lawmakers that welfare recipients should be guaranteed an annual income of $5,000 for a family of four. "I'm hungry," Representative Abner Mikva said. The Illinois Democrat said that "You're constantly aware of the fact that you're not getting enough to eat." The wife of Republican representative Paul McCloskey said that their fourteen-year-old son John was "simply starving" and that they were barely getting by on peanut butter, oatmeal, and grits. Representative Mikva had one hope for the project. "I hope it will make people aware that there are people who—on the officially sanctioned welfare budget—cannot get enough to eat."[10]

July 3: UCLA announced the imminent release of ARPANET, the Advanced Research Project Agency Network. ARPANET would be the forerunner of the internet, an early method of connecting computers together to share data and communicate. As the announcement prophesized, "We will probably see the spread of computer utilities which, like present electric and telephone utilities, will service individual homes."[11] The first test of connecting two computers together via the system was held on October 29, 1969, but it immediately crashed when the first two letters of the system's password were typed in.

July 3: Drama critic Clive Barnes said of the play, *Oh! Calcutta!* "[It] is the kind of show to give pornography a dirty name." The play was a controversial avant-garde revue of sex-related musical sketches focusing on sexual mores. The play had a small costume budget since the entire cast was nude for most of the two hour production. The play opened on Broadway June 17, 1969, and ran through 1972. A revival in 1976 kept the play running for another twenty years, making it one of the longest-running shows in Broadway history. The play's press agent, Shirley Herz, said that on this date, one of the beautiful actresses from the play had to leave quickly for City Hall on some urgent business. The actress only had on a sheer see-through costume. Quickly, the press agent grabbed a robe and the pair ran off down the windy New York City streets. The actress made it to City Hall much faster than she thought she would. As they walked down the street, the wind blew open the robe. The pair were arrested and had a quick ride to the city jail.

II

JULY 6–12, 1969

All Engines Running

6

WE HAD OTHER THINGS TO KEEP US OCCUPIED

Scattered across the jungles of North Vietnam, Cambodia, Laos, and South Vietnam were prison camps that were literally hell on Earth. These camps were known by many names to the hundreds of Americans who had been captured and placed in them during the Vietnam War: Alcatraz, Briar Patch, the Hanoi Hilton. Each one had a unique name, but they had one thing in common: horrid living conditions such that "squalor" does not come close to capturing them. They were places where prisoners were either starved or fed rancid food, places where unspeakable torture was inflicted to coerce military information from them, places where they were beaten into submission so that they would admit that they were criminals, only to be paraded before cameras as propaganda tools to show that they were anti-American.

Many times, the cells the prisoners were kept in had no windows. That was what awaited POWs when they arrived at a prison called The Zoo. All of the windows of each cell had been sealed off with bricks, which led to mental starvation that would have driven normal people insane. But the unwritten code of Vietnam prisoners of war was to hold out. Hold out for as long as you could.

The Zoo earned its name because each locked cell door had just enough play in it so that prisoners could peek out and also allow their captors and the livestock roaming the grounds to look in, making them feel like animals on display.

At this point, POWs could not communicate with one another, and no news or personal correspondence was allowed to enter or leave the prison. In an effort to maintain morale and their sanity, and to show defiance to their captors, POWs learned to communicate with each other through a method called the "tap code." It was a means of keeping contact with actual humans while at the same time keeping each other aware of what was happening around them.[1]

As Apollo 11 made the first lunar landing, Navy Lieutenant Mike McGrath, held captive in The Zoo, heard a rumor spreading through the compound that the United States had landed on the moon.

"Someone got a package," McGrath said in a HistoryNet interview, "and there was a sugar packet with a picture of Neil Armstrong standing on the moon with the flag." The news spread quickly through the compound via the tap code. "It was one of the brightest days."[2]

Another prisoner, Air Force First Lieutenant John L. Boling, found out the news during a brutal marathon torture session where he was tied up with shackles, ropes, and handcuffs in such a way that he was in desperate pain and was on the verge of breaking his bones.

When the captors left him alone in the room, Boling managed to loosen his binds and scoot himself to a nearby desk. He picked up a handful of envelopes lying there, but his captors returned and ripped them from his hands. Before the letters were snatched away, he saw a postage stamp on one of the letters with the US "Man on the Moon" stamp on it.

A merciless beating followed. When he was finally returned to his cell, he began tapping on the wall, "We own the moon!" Boling said it was a tremendous morale boost for the men.

On the battlefield, the men and women serving in Vietnam were often barraged with propaganda spewed over the radio airwaves by Trinh Thi Ngo better known to American troops as Hanoi Hannah. Ngo would read carefully written scripts in an attempt to demoralize the troops and convince them that the war was immoral and they should lay their weapons down and go home. In late July 1969, Ngo used Apollo 11 as a weapon in her broadcasts.

"The United States may be able to put a man on the moon," she broadcast, "but they cannot defend their bases in Saigon.[3] No one needs to go to the moon to see craters. All they have to do is look at the countryside of Vietnam. You can send a man to the moon and back but not bring your troops back from 10,000 miles away."[4]

For real news, troops stationed in Vietnam received the news of the day via letters from home, the Armed Forces Radio Network (AFRN), or the military newspaper *Stars and Stripes*. Peter Meloro had just arrived in Vietnam for his tour of duty only ten days prior to the moon landing.

"I was newly assigned to the 101st Airborne Division as a company clerk," Meloro recalls. "We did hear about the landing on [AFRN]. Quite frankly, it was somewhat of a non-event for me and I never sensed the event generated much discussion among the infantry troops. We had other things to keep us occupied."[5]

Indeed, they did. It would take another six years for the war to come to an end.

7

T MINUS TEN DAYS

During his short term in office, President John F. Kennedy made several overtures to the Soviet Union suggesting that both superpowers combine their space exploration efforts in the name of peace and as a way to share the cost of such an expensive venture. The first suggestion was tucked away within Kennedy's first address to a joint session of Congress. Embedded in the soaring speech rallying Americans to a better future and warning of the difficulties that lay ahead, Kennedy said:

> Today, this country is ahead in the science and technology of space, while the Soviet Union is ahead in the capacity to lift large vehicles into orbit. Both nations would help themselves as well as other nations by removing these endeavors from the bitter and wasteful competition of the Cold War. The United States would be willing to join with the Soviet Union . . . in a greater effort to make the fruits of this new knowledge available to all.[1]

Nothing of significance came from Kennedy's first or subsequent invitations, but as America was preparing to end the space race once and for all in July 1969, there was a glimmer of hope for cooperation between the countries.

The commander of the historic Apollo 8 flight that first orbited men around the moon, Frank Borman, had been dispatched by President Nixon on a worldwide goodwill tour during which one of his stops would be Russia. After laying a wreath at the tombs of Lenin, rocket scientist Sergei Korolev, and cosmonauts Yuri Gagarin and Vladimir Komarov, Borman and his family were escorted to the home of Russian cosmonauts, Star City, where he did a presentation on the flight of Apollo 8.

After all of the formalities of his visit concluded, Borman and his family traveled to Moscow on July 9, where he met with the president of the Soviet Academy of Sciences, Mstislav Keldysh, which was then followed up with a short forty-minute meeting with Soviet president Nikolay Podgorny. These meetings included precursory talks about possible joint US/USSR space missions. Following the meetings, Borman was greeted by reporters, where he told the gathered press that the meetings were "encouraging and beneficial."

"I consider that we should stop unnecessary duplication in planetary exploration," he said. "I would like to believe that in the not distant time when scientific laboratories will be in orbit, scientists of different countries in the world will cooperate aboard those ships."[2]

Borman's prognostic statement would become reality with the flight of the Apollo/Soyuz Test Project (ASTP) during which three Apollo astronauts docked with an orbiting Soviet spacecraft in 1975 and later with the construction of the International Space Station (ISS) in 2011.

When Borman returned home to the United States and was greeted by the press, he was quoted as saying that he was convinced that the Soviets were planning their own moon landing soon, not realizing that the Russian moon rocket, the N1, had exploded earlier in the month, dashing all hopes that the Russians could beat the Americans to the moon. But he did go on to say that the flight of Apollo 11 was all anyone in the Soviet Union wanted to talk about. "From people on the subways to their president, all I heard was that they are wishing success for Apollo 11."

The United Nations had a say in the Apollo 11 mission when they released a statement on July 11 reinforcing the view that there was no question that it was perfectly legal for the Apollo 11 crew to land on and explore the moon as long as it follows the guidelines set forth by the 1967 Outer Space Treaty that recognizes any country's right to explore space and the planets as "envoys of peace."

The statement went on to remind NASA and the astronauts that the treaty specifically forbids contaminating the moon with Earth germs (and vice versa), that they must provide full details of their findings to other countries, and that they must confine their activities to "peaceful pursuits."[3]

In Florida, the press had just received copies of the official 254-page Apollo 11 press kit that outlined every step of the upcoming flight. All seemed ready for the July 16 launch, but behind the scenes, there were still a few issues to iron out.

The first was how the landing and first steps on the moon would be televised. Even though Apollo 10's practice flight to the moon to test the lunar lander before the first actual moon landing had color television cameras onboard and proved to be highly successful, a debate still raged on about whether Apollo 11 should transmit the landing in color or black-and-white.

Engineer Maxime Faget argued that black-and-white images would be grainy and almost unrecognizable, but his argument fell on deaf ears as the Apollo Spacecraft Program manager, George Low, made the final decision—it would be black-and-white. Low cited scientists who said that black-and-white had higher resolution than color and with no atmosphere on the moon, color might not turn out at all.[4] Faget voiced his opposition but to no avail. He finally relented but put in one final jab at the decision saying that it was "almost unbelievable" that the culmination of a $20 billion program, and the most historic adventure mankind had ever taken would be "recorded in such a stingy manner."[5]

Of a more serious nature was the discovery of an increasing number of human errors turning up in the flight hardware, including inside the command module, the astronauts' home away from home on their journey. While no major errors were discovered during the testing of the Apollo 11 hardware, George Low appointed a team to do a "walk-down" of all future spacecraft before they were to be delivered to the Kennedy Space Center. Low understood that it would not be possible to identify every error, but at least this predelivery inspection would greatly reduce any future problems.[6]

These issues were far from the mind of the Apollo 11 crew as they continued preparations for the launch. At launch pad 39A, they practiced launch escape procedures from the top of the towering Saturn V rocket where they would jump into baskets that would send the crew zooming down a wire some 443 feet to the ground where an armored transport vehicle would whisk them away from the launch pad to an underground bunker 2,200 feet away. In the event of an emergency and the possible explosion of the rocket on the pad, the crew would remain in the bunker until it was all clear.

While Apollo 11 was on track for its meeting with destiny, NASA was also preparing for another historic event that would occur at the end of the month while at the same time facing the public spotlight and condemnation over another mission.

As engineers set their sights on the moon, two other missions were already winging their way deep into space heading for a rendezvous with Mars. They were the twin unmanned spacecraft, Mariner VI and Mariner VII. If all went well, the probes would pull off another space feat—flybys of Mars by two spacecraft one after the other. Launched from the Cape on February 24 and March 27 respectively, the spacecraft were reported to be in excellent condition as they sped across the blackness of space for the first dual mission to Mars. One of the probes would fly by the Martian equator; the other would fly past its south pole

Mariner VII was launched in March 1969 and would join its sister, Mariner VI, for a flyby past Mars at the end of July 1969. *NASA*

all the while snapping hundreds of photos of the Red Planet before whizzing off into space.

Having two interstellar probes flying a mission at the same time as the Apollo 11 moon landing posed some consternation with engineers. Both would require use of the big dish antenna in Goldstone, California, to receive telemetry data and photographs from the Mariners while at the same time picking up data, communications, and television images that would be transmitted back to Earth by Armstrong and Aldrin while they were on the moon.

Engineers set about the task of testing the Goldstone antenna for the cascade of information it would be receiving and determined that it would be able to handle both operations, but in the event of something going wrong, a backup plan was concocted where the lunar television coverage from *Eagle* would be handled like a classic baseball double-play, "Tinkers to Evers to Chance," only this time, it would be a complicated dance sending the signal from the Parkes receiving dish in Sydney, Australia, to a receiving station in the Pacific Ocean via microwave, to the United States via the orbiting Intelsat III satellite, to mission control in Houston, then to the television networks.[7]

The one major defeat NASA experienced this week came during what would be the final US flight of a live monkey being used as an experimental passenger shot into space aboard what was called a Biosatellite, or Biosat. Biosat was a series of missions that were launched into space to study the effects of weightlessness and cosmic radiation. The satellites that orbited the Earth contained a variety of test subjects ranging from plants to mice to primates.

On June 28, 1969, a fourteen-pound pig-tail monkey named Bonny was attached to numerous sensors, placed inside the tiny Biosatellite III capsule, and launched from Cape Kennedy atop a Delta rocket. The primary goal, as were the other flights of the Biosatellite series, was to explore the long-term effects of weightlessness on a body over a scheduled thirty-one-day mission.

Bonny was reported to be in good health on July 5, but soon after, the sensors attached to Bonny showed that her heart rate and respiration were decreasing rapidly. It was noted that she was sleeping more and for longer periods of time and she was refusing to eat, so much so that she lost 20 percent of her body weight. On July 7, after only nine days into her thirty-day flight, the decision was made to terminate the mission and bring Bonny home. Eleven hours after recovery, it was reported that Bonny died from a heart attack.

Reaction to the news of Bonny's death was swift. The non-profit United Action for Animals published a report documenting the inhumane treatment of the monkey, who had ten hair-like steel wires imbedded in its brain for the mission.

It would be the final flight of a primate into space by the United States, but sadly, the practice of using animals for experimentation continued unabated in laboratories and aboard rockets around the world.

One month prior to July 1969, the US Air Force had canceled its own ambitious space program, the Manned Orbiting Laboratory or MOL. An offspring of an earlier Air Force project, Project Dyna-Soar, which was canceled in 1963, the MOL picked up where Dyna-Soar had left off and built upon its technology.

The MOL was a joint project of the Air Force and the National Reconnaissance Office (NRO). The goal of the partially secret project was to loft two astronauts into space aboard a modified NASA two-man Gemini capsule, the same capsule that paved the way for the Apollo program by ironing out all of the mechanical details that would be needed on a flight to the moon. The capsule would be connected via a tunnel to a long tubular aft section that served as an orbiting laboratory. Each crew would work in space for two weeks.

The project was truly "partially secret." One part was the public-facing side, which officials said would study the effects of putting a person into space for an extended period of time. The classified part was to put a crewed spacecraft into orbit to spy on the Soviet Union.

In November 1966, a mockup of the capsule and laboratory was launched into space from Cape Kennedy's Launch Complex 40. Three years later, after spending $1.56 million ($130 billion in 2022 dollars), the program was canceled. Upon hearing the announcement of the cancellation, on July 2, the chairman of the Subcommittee on Space Science Applications, Congressman Joseph Karth, wrote to the Department of Defense asking if the technology developed to that point could be used in NASA's Earth Resources Satellite program. "It seems to me," Congressman Karth wrote, "that certain equipment, subsystems, sensors or techniques developed by the Department of Defense might be of use to NASA. May I have your advice on this matter?"

The reply came in the form of a government DD-173 form marked confidential. "The Manned Orbiting Laboratory (MOL) Program (USAF Program 632A)," the typed reply said, "was cancelled on 10 Jun 1969. Hardware material and information on the MOL program is not declassified or downgraded because the program has been cancelled. All commands are reminded that security requirements for MOL remain in force until revised or deleted through appropriate channels."[8]

Basically, the answer was "no."

As launch day drew near, in Washington, Representative John V. Tunney (D-California) brought to the floor of the House of Representatives a bill that would designate the date of the moon landing as "Space Exploration Day." Or "Lunar Day." Or "Moon Day." Whatever the final name would be, the day would be a national holiday.[9] On the senate side, Senators Spessard L. Holland (D-Florida) and Senator Edward J. Gurney (R-Florida) introduced bill S.J.R. 133 to revert the name of "Cape Kennedy" back to its original "Cape Canaveral" designation. The bill was referred to the Senate Committee on Interior and

Insular Affairs, but no action was taken on it. In October 1973, the Florida state legislature passed similar legislation, which restored the four-hundred-year-old name.

It wasn't only NASA, its contractors, and the crew of Apollo 11 preparing for the launch. Over two hundred professional astronomers from around the world and one amateur were getting their telescopes ready for the mission as well. The amateur was David Moore, a National Geographic Society cartographer. Moore was invited to be a "backyard stargazer" for NASA after he was one of the few stargazers to witness Apollo 10 on its return trip to Earth. What he had witnessed from his home in Wheaton, Maryland, were small bright flashes of light. Whenever the Apollo 10 spacecraft, *Charlie Brown*, would eject water from the service module into space, the water would instantly freeze and reflected sunlight.[10]

Even famed composer and bandleader Duke Ellington had a role to play in the flight of Apollo 11. On July 11 it was reported that the American Broadcasting Company (ABC) had commissioned "The Duke" to compose and perform an original score to commemorate the first moon landing.

The composition would be ten minutes long with Ellington singing lead on a song titled "Moon Maid." The performance of the piece would occur between July 20 and July 21 when Neil Armstrong would first step onto the moon.[11]

Another element of the upcoming launch of Apollo 11, one that was far from the minds of the crew, was the electricity that the launch was creating in the seaside villages of Cocoa Beach and Titusville. Five days ahead of launch, thousands of people began making their way into the towns to get the best viewing locations along the beaches surrounding the Kennedy Space Center. Tents began popping up along the area's causeways and the streets were beginning to gridlock. The *New York Times* described the scene:

> The temperature hovers around 100 degrees, leaving the ocean and swimming pools lukewarm. The air-cooled motel lobbies on Highway A-1A are a jumble of perspiring bellhops, yelling children, impatient parents, ringing phones, and the hard laughter of men in madras jackets who grip plastic glasses of Bloody Mary's at mid-morning.[12]

The head of the Cape Kennedy Convention and Visitor Bureau, Charles R. Johnson, was giddy about the scene. "Hell! This is it!" he shouted at reporters. "After 14 years it's here. It's almost as if I were a kid again."[13]

Stores all across Titusville, the Cape, and Cocoa Beach were now briming with Apollo 11 toys and T-shirts. The Associated Press ran a wire photo showing "pretty 17-year-old Cape Kennedy salesgirl Linda Grelle" showing off some of the hundreds of commemorative mementoes that were for sale at the space center to visitors who were coming in for the launch.[14] Menu items at restaurants all had Apollo- and moon-themed names, and bars were serving up Lift Off martinis for $1.25. Residents were warned about possible food short-ages due to the influx of tourists, and over a thousand police officers were called in to help control the crowd that was expected to exceed one million.

As the week was coming to an end, the flight of Apollo 11 officially began, not with the incredible ignition, roar, and rumble of the five massive Saturn V engines, but quietly and almost unnoticed as a large display countdown clock located in a viewing area along Banana Creek only 3.5 miles away from the rocket came to life at precisely 8 p.m. Eastern Daylight Time. On July 10, the large digital numerals glowed the news—liftoff was only ninety-three hours away. Including built-in holds in the countdown that would allow engineers to run through a series of prelaunch tests to determine if everything was still a "go" for liftoff and to give them an opportunity to correct any sort of gremlin that may rear its ugly head, the epic voyage would begin in five days.

8

I WANT TO FORGET
THIS AWFUL PLACE

An airliner taxis to the gate at a nondescript airport and stops. As the door of the aircraft opens, four hundred young men begin to file out and step onto the tarmac. On the other side of the gate, hundreds of other young men wait to board another plane. They begin shouting at the new arrivals: "Remember to duck!" "Charlie's waiting for you!" "I Hope you brought your steel underwear!"[1]

Suddenly from out of nowhere, the new arrivals are greeted by the explosion of incoming rockets being fired on their location scattering the men into sandbagged bunkers. The attack doesn't last long, but it is enough to put them on edge.

"Odors are intense [in Vietnam]," Michael Zboray writes in his memoir about his tour of duty in Vietnam. "The most offensive smells are those of burning or decaying human flesh. If you've ever felt like the guy sitting next to you on the bus was a little smelly, then try to imagine how disgusting the human body gets."[2]

In relating his tale, Zboray makes a statement that is echoed by many Vietnam War veterans: "I want to forget this awful place. Yet, at the same time, I never want to forget the many lessons that enabled me to mature quickly from just a skinny kid with a scrabbly mustache, into a young man working so hard to seem older than his few years as he evolves into manhood."[3]

During the early years of United States involvement in Vietnam in 1953 and up until 1964, most of the Americans stationed in the country were advisors, clerks, and military personnel who were aiding the South Vietnamese government and military in their fight against North Vietnamese aggression. In that first year, only 760 US troops were in the country. Those numbers, however, were about to change dramatically. Another statistic that was about to explode was the US death count in the country. Records from the National Archives show that for the first eleven years of US involvement, reported US deaths amounted to a total of 416.[4] Only five years later in July 1969, the number had skyrocketed to almost 40,000.

Following the assassination of President Kennedy in 1963, Lyndon Johnson was sworn in as the country's thirty-sixth president. When Johnson ran for reelection, he campaigned as the peace candidate, vowing that he would not escalate the conflict in Vietnam, unlike his Republican challenger, Barry Goldwater.

Almost a year after the election in late 1964, and in response to a reported attack by the North Vietnamese on US warships in the Gulf of Tonkin, the death toll would begin to rise exponentially as Johnson ordered a campaign of sustained bombing over North Vietnam and of the Ho Chi Minh Trail, a series of interconnecting routes that ran from North to South Vietnam via the countries of Laos and Cambodia. It was called Operation Rolling Thunder. It was Johnson's belief that the operation would force the North Vietnamese into submission and they would recognize the sovereignty of the South, thus giving up all ideas of joining the two countries together under Communist rule. It was during this operation that the first battalions of US Marines arrived in the country, and by the end of 1965, over thirty-eight thousand soldiers had landed.

By the end of 1968, the number of soldiers and advisors stationed in Vietnam was topping 535,000. The number of American casualties neared 40,000. That year, America's most trusted journalist, Walter Cronkite, was a believer in the Vietnam conflict and that the US government could stop Communist North Vietnam from taking control of the South through might and a decisive victory.

Cronkite was a veteran war journalist having won the hearts of Americans with his coverage of battles during World War II. Now in 1968, annoyed by the attitude being displayed by younger journalists covering the front lines with their cynicism toward the conflict,[5] Cronkite paid a visit to the frontline himself and devoted an airing of the *CBS Evening News* to his findings, which had changed dramatically.

"It seems now more certain than ever," the seasoned journalist concluded, "that the bloody experience of Vietnam is to end in a stalemate. . . . It is increasingly clear to this reporter that the only rational way out then will be to negotiate, not as victors, but as an honorable people who lived up to their pledge to defend democracy, and did the best they could."[6]

Public sentiment against the war, particularly with the youth of America, had already turned, and their dissatisfaction and the call to end the war was spilling out into the streets. With Cronkite's words, the media had joined the protest. The Cronkite editorial changed how television news media covered the war. It became known as the Television War as every night, the networks would bring the death and destruction directly into homes across America. Many of the network news shows began ending their broadcasts with the body count for the day and the total deaths up to that point of the war.

In a legendary piece of American history, Johnson reportedly told aides that if he had lost Cronkite, he had lost the war. Johnson made a primetime speech announcing that he would not run for reelection, opening the door for more anti-war leaning candidates to fill the void.[7]

Following both the assassination of Robert Kennedy while on the presidential campaign trail in Los Angeles and the violence of the Democratic National Convention in Chicago, voters rallied around Richard Nixon and elected him the country's thirty-seventh president.

Only weeks after taking office, Nixon directed the military to increase the bombings with the same hope Johnson had—push the North Vietnamese into submission. In an off-the-cuff interview with the press on a trip to the island of Guam in July 1969, Nixon told reporters that the United States could no longer afford to defend its allies completely. The American military was being stretched to its limit and was the biggest line item in the US budget. The United States would continue to honor their previous agreements around the world, he said, but it was time to pull back on resources.

While the comments were not directed solely at the Vietnam War, they did give the president a chance to placate the American public. Shortly after the remarks were made, Nixon began a policy of "Vietnamization" in which the United States would aid the South Vietnamese with artillery and weapons and teach them how to use those weapons to defend themselves. They would also instruct the South Vietnamese troops on combat strategies while at the same time withdrawing troops, which had burgeoned to over 535,000.

Behind closed doors, Nixon told senators on Capitol Hill that he was concerned over increasing public opposition to the war and what the political consequences of that would be. Without taking some sort of public-facing action, he feared that the war would jeopardize Republican chances in the next midterm election. With that in mind, Nixon began advocating for a nearly complete withdrawal of troops by the end of 1970, and in a move that he believed would placate the public, he publicly announced that he would be withdrawing twenty-five thousand Americans from the country as part of the first wave of his Vietnamization strategy.

On the morning of July 8, a contingent of 814 members of the Third Battalion, Sixtieth Regiment of the Ninth Infantry Division had their duffel bags packed and waited on the tarmac of the Tan Son Nhut Air Base near Saigon for the long trip home. These would be the first Vietnam veterans to return home from the war. The next two days, however, would be trying for the men who just wanted to get home. US, Vietnamese, and local government officials had something else in mind—ceremonies. Lots of ceremonies.

The battalion that was assembled on the tarmac had to endure two hours of military fanfare including marches played by local and military bands and a seemingly endless array of speeches from a plethora of military officials including the commander of American forces in Vietnam, General Creighton W. Anderson, South Vietnamese president Nguyen Van Thieu, and South Vietnamese defense minister Nguyen Van Vy.

"Together we have repelled Communist aggression," Thieu told the soldiers whose division had lost 1,855 members. "Our duty is to make sure their sacrifice was not in vain."

When the fanfare had finally subsided, the men boarded a plane bound for McChord Air Force Base in Tacoma, Washington. As each man boarded the plane, they were greeted by young Vietnamese girls who presented them with a small gift and a lei.

As the first plane that was filled with newly minted Vietnam veterans touched down in Tacoma, marching bands began playing "Hands across the Seas" and "When the Caissons Go Rolling Along." Army Chief of Staff General William Westmoreland greeted the men saying, "You can stand tall and proud. You can look any man in the eye knowing that you have served your country when you were called."[8]

The following day, after spending a year in the jungles of Vietnam with death and carnage all around, the men were expected to relearn basic parade drills before they could go home. The troops spent the day marching up and down a parade ground, spinning on their heels and relearning how to march in unison all in preparation for what followed on July 10—a celebration of the first men returning home after Nixon's withdrawal strategy was put into place.

As the men marched down the windswept streets of Tacoma, they were greeted with fields of American flags waving in the breeze, tickertape snowing down on them, and signs that ran the gamut of emotions from a simple "Thank You!" to "We'll Stay in the Streets Until All the GIs Are Home" and "Win in Vietnam. U.S. Forever. Surrender Never."

The marching band music was accentuated with cheers of love and support that were occasionally interrupted with shouts of "Go home, Commies" from a small group of protestors. While the protestors were outnumbered many times over by those in attendance who were appreciative of what these young men had been through and the sacrifices they had made, everyone in the crowd exuded the same feeling—welcome home, boys, but let's bring them all home now.

The irony of the moment is that as these 800 men from the first contingent of the 25,000 troops Nixon promised to bring home were being celebrated, an additional 1,000 men were hopping aboard Air Force C-141 aircraft to begin their own tour of duty in Vietnam, with another 10,000 to follow suit.[9] That same day it was announced that an additional 2,157 American soldiers had died and many more were returning injured.

For some soldiers returning home, the homecoming was not as glorious. The injured would return on stretchers, where they would be immediately transferred from the plane to a waiting bus. As CBS News correspondent Charles Kuralt said of the scene, it was "institutionalized compassion. They were handled with gentleness, but on a production line."[10]

Once on the bus, they were greeted by a steely-eyed officer who barked instructions in rapid succession.

> Gentleman, may I have your attention for a couple of minutes, please. I believe I can answer some questions for all of you. These remarks are primarily for those of you returning from the line of fire. . . . When you left the live fire area, most of you left behind most if not all of your gear. Please bear in mind, gentlemen, that the only way old Uncle Sam can repay you for your personal property which has been lost is if you submit a claim for reimbursement. . . . Now the $64 question is what's the score on convulsant leave?
>
> This is a free ride, not chargeable to you in any way. So as soon as you can convince the doctors and the nurses at the station you are going to that you're healthy, you're full of fun and games, that you can take care of yourself and that you will be no problem to your dependents, you will have to convince the doctor that you can stand the wear and tear of thirty days leave on the economy.[11]

When it came to anti-war protests in the United States, there were two mindsets. The first was that the war was unnecessary and was killing our young people needlessly. The other was a show of opposition toward the means in which the military kept building up its ranks through the military draft or what is also called conscription.

The military draft required that all male citizens and immigrants ages eighteen through twenty-five register with the government. From that list, men

'MY LAST EMPLOYMENT? — VIETNAM'

Returning Vietnam veterans came home and were increasingly hard-pressed to find work in civilian life. Many had no skills or job experience after being drafted or enlisting in the military straight out of high school. Times were particularly rough for returning minority soldiers. *Drawing by Edmund S. Valtman, "My last employment?—Vietnam." United States, 1971. Library of Congress. https://www.loc/gov/item/2016687293/*

would be called into service through a random lottery. The draft of the late 1950s and early 1960s was the product of the Selective Training and Service Act of 1940 that was signed into law by President Franklin D. Roosevelt in anticipation of the United States entering World War II. During the Korean War, the act was modified so that college students who were in the top half of their class grade wise could defer their induction into military service.

Following the end of the Korean War in 1954 through 1964, the draft was considered a "peacetime" draft and was used to maintain a quota of military personnel during the Cold War with the Soviet Union. During this period, many college campuses began requiring that their male students begin mandatory enrollment in ROTC, which fueled anger among the students, and the first wave of protests against the draft began.

With President Johnson's escalation of the Vietnam War in 1965 came the need to build up troop strength in Southeast Asia. The number of American men drafted into service jumped significantly from 1.4 million during the entire ten-year peacetime draft to over 2.2 million during the Vietnam War between 1964 and 1973. That's over 300,000 men per month.[12]

Protests to the draft ranged from full-blown, take-to-the-street marches to individual acts of defiance. On June 14, 1968, the world's most famous baby doctor, Dr. Benjamin Spock (the author of the bestselling book, *The Common-sense Book of Baby and Child Care*), along with Yale University Chaplin William Sloane Coffin Jr. were found guilty in a federal court of "conspiring to aid and abet registrants to violate the Selective Service Act." The case stemmed from Spock's participation in an April 1967 anti-war march of an estimated three hundred thousand to the United Nations building in New York that was led by Dr. Martin Luther King Jr. and entertainer Harry Belafonte. Spock was one of the original signers of the Call to Resist Illegitimate Authority manifesto that supported draft resistance and the right of young men to disobey "illegal and immoral orders."[13]

On July 24, 1969, the draft conspiracy convictions of Spock and three other conscientious objectors were thrown out by the First Circuit Court of Appeals. In the ruling, Dr. Spock was acquitted of all charges. Coffin was not acquitted but was granted a retrial due to the prosecution being prejudiced in the original trial.

"The tragedies," Spock told the Associated Press following the verdict, "are that the war is still dragging on and that young men have been imprisoned for being opposed to it and doing as their consciences dictated."[14] Spock added that he hoped that "100,000, 200,000, or even 500,000 young Americans either refuse to be drafted or to obey orders if in military service." Dr. Spock went on to participate in several more anti-war protests that year and continued advising men about opposing the draft.

Another notable draft case in July 1969 centered around the heavyweight boxing champion of the world, Muhammad Ali, and his refusal to be inducted into the Army. On July 24, Ali appeared in a federal courthouse and was convicted of avoiding the draft.

The incident began on April 28, 1967, when Ali arrived at an induction center in Houston, Texas. While some conscientious objectors fled to Canada or burned their draft cards in the street as a show of defiance, Ali refused to be sworn in. It should not have come as a surprise to anyone because two years earlier, Ali was quoted as saying, "My conscience won't let me go shoot my brother, or some darker people, or some poor hungry people in the mud for big powerful America. And shoot them for what? They never called me nigger, they never lynched me, they didn't put no dogs on me, they didn't rob me of my nationality, rape and kill my mother and father. . . . Shoot them for what? How can I shoot them poor people? Just take me to jail."[15]

For his actions, Ali was stripped of his boxing license and title. He was allowed to return to the ring in 1970 and his conviction was overturned by the US Supreme Court in 1971.

While young men were working out ways to defy the draft, a group of women in Manhattan tried a different approach. On July 1, five women including Kathy Czarnik from Detroit, Michigan, broke into the Forty-Fourth Street draft board where they ripped phone wires out of the walls, destroyed typewriters, and shredded over two thousand draft records. Before leaving, the women pasted the walls with photos of dead soldiers that had appeared in *Life* magazine.[16] The women were able to leave the office without detection but then appeared two days later at an anti-war rally at Rockefeller Center where they publicly admitted to the break in.[17] As they spoke, shredded paper began to fall. It was the shredded draft files.[18]

FBI agents monitoring the event quickly moved in. A fight between the agents and some members of the protestors broke out but, in the end, the women were arrested and placed under a $2,500 bond.

Anti-war protests took many forms and were not only relegated to the youth of the world. On July 7, as a group from the American Friends Service Committee, a group of people made up of the Quaker faith, held a quiet sit-in outside the White House holding up signs that read, "Quaker Meeting for Worship, for Peace in Vietnam," five of its members were inside the building holding meetings with the nation's highest profile Quaker, President Nixon, and Secretary of State Henry Kissinger to discuss the administration's Vietnam policy.[19] Nothing came of the meeting. Later in the year, that same group ran a large ad in the *New York Times* that simply read, "You cannot make peace by making war."[20]

In New York City, a call went out for a massive weeklong protest that was called Hiroshima Week. The ad in the Village Voice promoting the event headlined, "Moonstruck? GI's Want to Land in the U.S." The ad went on to list all of the events planned around the anniversary of the dropping of an atomic bomb on Hiroshima, Japan, in World War II. The ad focused on the disparity of landing on the moon while thousands of men were dying in Vietnam: "It's kind of hard for a guy to get too excited about the out-of-this-world accomplishment of Apollo 11 when he's stuck in the jungles of Vietnam and all he can think of is one thing—landing back in America."[21]

Events held during Hiroshima Week would be staged across the city, including speeches at various churches, demonstrations at city council meetings, candlelight vigils, and the reading of the names of all of the war dead.

While most of the country, and the world, wanted US involvement to end in Vietnam as quickly as possible, preferably with just a complete withdrawal, some were advocating for complete victory, including the Young Republican National Federation (YRNF) who met on July 11 in Chicago. In a resolution adopted by those in attendance, the YRNF urged President Nixon to "use all powers at his command to successfully and honorably terminate the conflict." Later in the session, the resolution was modified to include the words "through victory."[22]

It would take another six years for the Vietnam War to come to an end as the final troops were withdrawn and North Vietnamese tanks rolled through the gates of the Presidential Palace in Saigon. In the end, 2.7 million served during the conflict. Of them, 58,220 had died with 304,000 injured, not to mention the trauma and long-lasting effects the war had on the men who served during the conflict, a trauma that time can never erase.

9

THE GREATEST THREAT TO THE INTERNAL SECURITY OF THE COUNTRY

From May 1924 to May 1972, the Federal Bureau of Investigation (FBI) was led by one of the most successful and, arguably, most controversial, directors, John Edgar Hoover.

After obtaining a law degree from George Washington University Law School in 1916, Hoover began his career in law enforcement in 1917 with a position at the Department of Justice. From there, the twenty-nine-year-old's ambitions continually drove him forward so that in only seven short years, he had reached the pinnacle of his career, the director of the FBI, in 1924.[1] That career saw Hoover serve as director under every president from Calvin Coolidge to Richard Nixon.

During his early career, Hoover popularized the image that FBI agents, or *G Men*, were gun-toting avengers, taking down communist labor agitators, kidnappers, mobsters, and bank robbers. That image was buttressed with infamous shootouts between agents and the likes of John Dillinger and Lester Gillis (a.k.a. "Baby Face" Nelson) in 1934.

Hoover would often direct his agents to infiltrate and dismantle "leftist" union and political organizations by unique and often dubious methods including burglary, inciting criminal activity, secret wiretaps, and planting false evidence. In a 1965 Gallup Poll, 85 percent of the country saw Hoover and the agency favorably. By 1969 when Hoover's tactics were becoming increasingly obvious, the agency's popularity had dropped, with only half the country giving them a favorable rating.[2]

In 1956, Hoover initiated the Counterintelligence Program (COINTEL-PRO) as a method of disrupting the activities of "political radicals" in the

United States, in particular, the Communist Party. In the early 1960s, the program's mission was expanded to include the Ku Klux Klan, and by 1969, the program also targeted the Black Panther movement.[3]

The Black Panther Party for Self Defense was founded only a few short years prior in 1966 when its organizers, Bobby Seale and Huey P. Newton, met at Merritt College in Oakland, California. What made the Panthers or BPP different than the Southern Christian Leadership Conference that was led by the late Dr. Martin Luther King Jr. is that King promoted change through nonviolent means, while the BPP believed in Black nationalism, socialism, and armed self-defense in cases of police brutality.[4] The name was inspired by a logo used by a lesser-known activist group from Lowndes County, Alabama, the Lowndes County Freedom Organization.

The foundation of the Black Panthers was their Ten Point Program:

1. We want freedom. We want power to determine the destiny of our Black Community.
2. We want full employment for our people.
3. We want an end to the robbery by the Capitalists of our Black Community.
4. We want decent housing, fit for shelter of human beings.
5. We want education for our people that exposes the true nature of this decadent American society. We want education that teaches us our true history and our role in the present day society.
6. We want all Black men to be exempt from military service.
7. We want an immediate end to POLICE BRUTALITY and MURDER of Black people.
8. We want freedom for all Black men held in federal, state, county and city prisons and jails.
9. We want all Black people when brought to trial to be tried in court by a jury of their peer group or people from their Black Communities, as defined by the Constitution of the United States.
10. We want land, bread, housing, education, clothing, justice and peace.

Despite the communist and revolutionary elements of their plan, and their armed defense against police brutality, the Black Panthers had another side to them, one that provided much-needed social programs to Black communities such as legal aid, adult education classes, and screening for sickle cell anemia. In January 1969, a new program was initiated by the Panthers, Free Breakfast for School Children. Party leaders and volunteers would solicit donations from local grocery stores of nutritious breakfast foods including eggs, cereal, milk, and fruit and then prepare the meals and serve them for free to the children.

The program was first tried out at an Episcopal school in Oakland. On the first day, eleven children enjoyed breakfast, many of whom had never had breakfast before. Teachers were thrilled with the results. A church parishioner marveled that the children "weren't falling asleep in class [and] they weren't crying from stomach cramps."[5]

What started as a small experiment in Oakland quickly gained in popularity, and soon, thirty-six similar programs were established across the country feeding tens of thousands of school children.

To Hoover, however, there was something nefarious about this generosity. The BPP was already on the FBI's radar for their posture on armed self-defense against police as well as the party's communist and revolutionary elements. In his annual fiscal report released in June 1969, Hoover set the stage for what was to come, stating that there was an increase in activity among violence-prone Black extremist groups. "Of these," the report said, "the Black Panther Party, without question, represents the greatest threat to the internal security of the country."[6]

"Leaders and representatives of the Black Panther Party," the report continued, "travel extensively all over the United States preaching their gospel of hate and violence, not only to ghetto residents, but to students in colleges, universities, and high schools as well" and Hoover vowed that 1969 would be the end of the Black Panther Party.

To Hoover, the free breakfast program was a way of grooming new radical members into the Black Panther Party, and he directed the FBI into action in an attempt to put an end to the program. In an FBI memo from early July 1969, an agent reported that the Indianapolis chapter of the BPP was attempting to establish a free breakfast program similar to the one in Oakland: "The Indianapolis chapter of the Black Panther Party has in recent weeks been collecting or attempting to collect funds for the purpose of starting a Breakfast for School Children Program. . . . It is felt by the Indianapolis division that a quick counterintelligence move might bring to a halt their attempts to establish a breakfast program."[7]

The memo went on to request permission to anonymously send literature that would denounce the program and the BPP, referencing several articles that could be disseminated among ministers of the area's Black churches to "persuade" them to halt the program.

A later memo from July 10 between the San Francisco bureau and New York Office (NYO) described similar counterintelligence methods that could be used

A Basic Contradiction in Affluent America:

Hunger in the Land of Plenty

CHICAGO YOUNGSTERS joke with photographer as they enjoy 7 a.m. breakfast provided by the Illinois Black Panther Pary in its breakfast for schoolchildren program which is being pushed in ghettos across the country.

By Skip Bossette
(Black Press International)

All across the nation, police have a 24-hour alert on for the Black Panthers. This alert has resulted in numerous attacks on the Panthers, the cold-blooded murder of some and the jailing of others—including Black Panther Party leader Huey P. Newton, currently being held in a California prison.

HOWEVER, frequent attacks on them in California, New York and other cities where Panther chapters have formed have failed to stop the young men and women who make up this new breed Black rights group from working to bring a better life to their people.

While police and establishment newspapers try to characterize the Panthers as "Black anarchists," the party is working hard to promote programs which are designed to help the "people of America" to take charge of their own destinies and free themselves from the blood-thirsty grip of the world exploiters.

"Power to the people" is a Black Panther by-word. It suggests that corrupt politicians, crooked police and blood-lusting businessmen should be divested of the reins of power. The people, say the Panthers, must be the masters of the world.

EVEN AS PANTHER leaders travel the country and other parts of the world, offering their help in organizing the inevitable struggle for government in the interest of people, their local chapters are busy building programs which seek to solve problems created by the current profit-responsive government of the United States.

It is no secret that the United States government could wipe out poverty in America if it wanted to do so. It is also no secret that the same government does little to make any attempt to ease the problem. Instead, little children are left to go hungry while fat businessmen divide up the spoils of the latest "defense" program and raise themselves pay raises.

The Panthers, unable to stop the Congressmen, have instituted a program which hopes to at least feed some of the hungry children—so that their aching bellies might be eased enough to allow them to get through a school day without being torn by the pains of hunger.

ALTHOUGH THE Black Panther breakfast for children plan is operating in many American car-

ried out by high school and college students who rise with the dawn and go to feed their hungry little brothers and sisters before continuing on to classes.

A driving rain greeted us when we left for Chicago's West Side Better Boy's Foundation, 1512 S. Pulaski. The cold rain beat bitterly against the windows showing rows of tables with scores of elementary school children seated hungrily around them.

Inside, Chicago chapter Deputy Minister of Labor Yvonne King —a lovely young woman who wore a purposeful look—explained that "the children are fed hot meals—toast, eggs, milk, beef bacon, oatmeal, grits and fruits." It was obvious that the children approved.

"THE PROGRAM was created by Black Panther Party headquarters in Berkeley, California," Miss King continued. "It was realized that there are many oppressed children going to school hungry throughout the country. This program was clearly one of the needs of the people and the Black Panther Party is here to serve the people's needs."

As she spoke, more children came in through the door—shaking rain off themselves and finding space at the tables where other Black Panther Party members were serving them. Despite the number of children present, the room was quiet except for good natured joking between the Panther members and the children who seemed to regard them as some kind of benevolent big brothers.

"Hunger among schoolchildren illustrates one of the basic contradictions in American society," Miss King pointed out. "America is one of the richest nations in the world, able to send countless numbers of rockets into space at the drop of a dollar, yet people are starving.

"WE ARE SHOWING the people that their children can be fed free."

Miss King estimated that the breakfast for schoolchildren program was currently feeding about 500 youngsters who otherwise might go to school with gnawing emptiness in their bellies. She said the program at the Better Boy's Foundation took care of about 225 per day while a program on the South Side, 500 E. 37th Street, took care of another 250 to 300 on Mondays, Wednesdays and Fridays.

"Funds come from various organizations and individuals; food

BLACK PANTHER PARTY member helps children keep table neat while they continue with pre-class breakfasts Panthers have initiated with the help of businessmen and people interested in the welfare of ghetto youngsters.

Black Panthers are Feeding Hungry Children in Ghetto

programs, we are trying to involve the community to the point where local people can take over. Our efforts are a people thing. They are for the people. The program is meant to be run by the people."

Despite obvious police determination to wipe the Black Panther Party off the American map, the party is gaining new members every day—on the streets of the ghetto and in colleges across the

"but more is needed. Party members are continually engaged in a search for new resources."

"IF WE COULD only get a dollar per week from each person who is concerned about the welfare of our beautiful Black children," another Panther spoke out, "our problem wouldn't be so big."

"In this program," Miss King went on, "as in most of our

land. Its leaders are becoming national heroes—for whites and Blacks alike—despite news media attempts to picture the Panthers as gunsels and terrorists.

The basic philosophy behind the Black Panther Party may frustrate every establishment attempt to destroy its members. "Power to the people!" the young men and women said as we left. "Power to the people!" we replied.

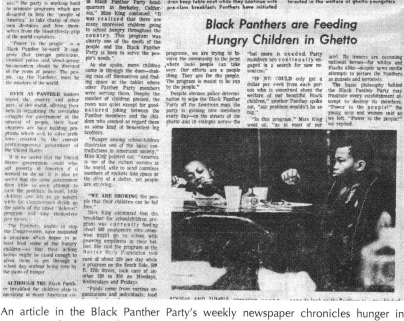

An article in the Black Panther Party's weekly newspaper chronicles hunger in poverty-stricken areas of America and the BPP's free-breakfast program for school children. *Internet Archive,* https://archive.org/details/03-no-2-1-22-may-4-1969/page/n11/mode/2up. *Black Panther article May 4, 1969, page 13*

to thwart efforts by the BPP to use the All Saints Roman Catholic Church in New York City as a site for a children's breakfast program. Once again, the NYO was directed to use anti-Panther literature that demonstrated the "true feelings and attitudes of the BPP to the people of the community" in order to dissuade the church from moving forward with the program; only this time, an FBI agent would personally deliver the message.[8] "Direct contact should be made through the established liaison source by a mature and experienced special agent who is familiar with various aspects of racial matters."

The memo also suggested that the agent making contact should stress that Harlem's NOI Mosque #7, which was located near the Catholic church in question, may "take exception to the facilities of a Roman Catholic church being used to dispense the 'White Man's food' to Black children and may consequently lead to violence in some sort to stop this." They were also to point out that the food donated for the breakfast program may have been obtained through extortion.[9]

The success of the Black Panther Party to feed impoverished children overwhelmed Hoover to the point where state and local governments took notice of their efforts and picked up and ran with the BPP program. One high-ranking US government official admitted that "the Panthers are feeding more kids than we are."[10] Cities like Chicago began diverting federal funding for other projects to their own local version of the Breakfast for Children program, and before long, the federal government's Child Nutrition Act was expanded to include free breakfasts and lunches to poor school children across the country.

J. Edgar Hoover's attempt to paint the Black Panthers as simply a group of thugs was thwarted, and his efforts to destroy the BPP's social program went down in flames. In 1971, all FBI counterintelligence operations were suspended after a strong backlash by Congress and the American public. The Panthers would continue to exist all the way through 1982, when the organization was officially dissolved.

The depth of Hoover and the FBI's surveillance of Americans ranged far beyond activist organizations. They also targeted individuals in the entertainment industry for their roles in assisting these organizations. Norma Jeane Mortenson, better known as actress Marilyn Monroe, was suspected by Hoover of being a communist sympathizer. Beatle John Lennon and his wife Yoko Ono were tracked for their anti–Vietnam War rhetoric. Even baseball legend Jackie Robinson was being investigated for his civil rights activism. In July 1969, the "Queen of Soul" Aretha Franklin was added to that list.

In a July 1969 memo titled "Possible Racial Violence, Urban Areas, Racial Matters,"[11] Franklin is identified as a person under investigation for her role in the 1960s civil rights movement.

The memo was part of an ongoing investigation into the singer that stemmed from an episode that occurred the previous year at the Red Rocks Amphitheater in Denver, Colorado. The singer was scheduled to perform at the venue but with only minutes before the concert started, Franklin learned that the promoter refused to pay her the $20,000 she was promised for the performance. Franklin refused to perform and the crowd that had been waiting patiently for the show to begin went into a rampage when they heard the news. They destroyed thousands of dollars' worth of sound equipment, chairs, and pianos and set fire to trees and trash. But the FBI surveillance of Franklin was not only based on the riot that ensued with the cancellation of the concert. They believed she was part of the militant Black power movement of the 1960s.

Additional memos released to *Rolling Stone* magazine in 2022[12] showed that the FBI was monitoring the singer for her performances at Southern Christian Leadership Conference (SCLC) and National Association for the Advancement of Colored People (NAACP) conventions in 1967 and 1968. Four days after Dr. Martin Luther King Jr.'s assassination, Franklin was scheduled to appear at a memorial for King along with Mahalia Jackson, the Supremes, and Sammy Davis Jr. "Of this group," one memo read, "some have supported the militant Black power concept and most have been in the forefront of various civil rights movements."[13]

The surveillance continued well into 1971 when the singer was believed by the FBI to be part of the Black Panther Party, which, as we have seen, was already under scrutiny by the agency, the Black Liberation Army, and the Boston Young Workers Liberation League, an up-and-coming communist party in the United States.

In 1973, two FBI informants told agents that they never knew Franklin to be associated with any radical movements, and after years of tracking her, in a final memo, the agency relented. "In view of the fact that there is no evidence of involvement by Miss Franklin in [Black Liberation Army] activities," the memo said, "and in view of her fame as a singer, it is felt that it would not be in the best interests of the Bureau to attempt to interview her."[14]

When it comes to the law and an alleged criminal's rights, there is one legal precedence that stands above the rest: the right to remain silent. This right was a landmark decision by the US Supreme Court in 1966 which ruled that a person's Fifth Amendment right restricts prosecutors from using a suspect's statements made in response to interrogation while they are in police custody unless there is ample evidence that the suspect was notified of their right to legal counsel and that statements they make can be used against them in court.

Miranda vs. Arizona stemmed from a March 1963 case in which Ernesto Miranda was arrested by the Phoenix police in connection with the rape of an eighteen-year-old woman. After two hours of interrogation, Miranda signed a statement, was charged with rape and kidnapping, and later sentenced to up to thirty years in prison. Miranda's attorney objected stating that his client's confession was garnered without his knowledge of his right to counsel. The appeal made it to the Supreme Court, where Miranda's conviction was overturned in a 5–4 decision.[15]

Since that time, it has become mandatory that if a suspect is brought in for questioning, they must first be notified of their rights against self-incrimination and the right to counsel. In July 1969, US Attorney General, John Mitchell had a different take on the law. In a memo released to the media on July 31, Mitchell stated that the Miranda Law should only be considered as one of the aspects to consider in determining if a suspect's confession was voluntary or not.

"In general," the memo said, "federal lawyers and law enforcement officials will continue the present practice of giving a suspect a full warning of his right to remain silent, of his right to an appointed or retained counsel, and of his right to terminate his questioning whenever he chooses. However, if a federal official inadvertently fails to give a full warning, the Department of Justice now believes that the confession may still be a voluntary confession and should be presented to the court as evidence."[16]

In a nutshell, what Mitchell was saying is that failure to give a full warning does not necessarily mean that the confession is invalid. Over the course of the Nixon administration, the rights of the accused established by Miranda were chipped away in particular by the four Supreme Court justices appointed by the president. While the law is "on the books," it is no longer used as strictly as it was originally intended.

During his presidential campaign in 1968, Nixon promised to change a set of laws that dealt with the nation's ever-growing drug problem. Seven months after his inauguration, on July 14, President Nixon proposed these new laws in a message to congress urging them to take strong action to confront what he called the "wretched trafficking" of drugs in the United States. These proposals later were called the War on Drugs.

In his message, Nixon cited an 800 percent rise in juvenile drug-related arrests between 1960 and 1967. "The number of narcotics addicts across the United States is now estimated to be in the hundreds of thousands," Nixon said before proposing a ten-step plan to thwart the ever-growing issue.

The most consequential of these proposed laws would be to give federal narcotics agents "no-knock" authority, where they could obtain a search warrant from a judge then enter a suspect's house by "breaking down the door, window, or any other part of the building" without knocking or identifying themselves.

During a White House briefing on the message, a spokesperson said that the no-knock law would allow officers to seize easily disposable drugs before they could be flushed down a toilet or burned. When the spokesman began taking questions from the press, a reporter asked, "If he [the officer] is dressed in civilian clothes and goes on in there, what happens if the man [suspect] shoots him?" To which the spokesman nonchalantly replied, "He will be tried for murder."[17]

Congress codified no-knock warrants soon after the president's request in the Drug Abuse Prevention and Control Act and the District of Columbia Court Reform and Criminal Procedure Act, and the president signed the procedure into law.[18] Over the next four years, stories of abuses of the law were rampant. The common thread—agents would break down doors and hold families at gun point as they ransacked their home before the officers realized that they were in the wrong home. In many cases, the agents did not receive a warrant to even conduct a raid.[19] A two-month investigation in 1973 in New York City found that four innocent people had been killed in one such raid. And that was only in two months.[20]

Senator Charles H. Percy had heard enough. The Republican senator from Illinois introduced a bill that would repeal the no-knock provision of the new drug laws and provide a means of reparation for the families who have fallen victim to these actions.

"If the past few months have taught us anything," the senator told reporters, "it is that excessive zeal, even in the pursuit of so worthy a purpose as drug law enforcement, cannot be allowed to destroy the fundamental rights of American citizens. . . . We cannot deny basic justice to the victims of the fast and loose tactics of a federally sponsored law enforcement program."[21]

The law was repealed in 1974, but its legacy continues to this day. As of 2022, thirty-six states allow the use of no-knock warrants; twelve restrict their use, with only Florida, Oregon, Connecticut, Tennessee, and Virginia outright banning the law. For those states that still allow no-knock warrants, the results have been the same as they were between 1969 and 1974.[22]

The highest profile case being the tragic killing of twenty-six-year-old emergency room technician Breonna Taylor in her home in Louisville, Kentucky, on March 13, 2020. Police officers had a warrant to search Taylor's apartment for illegal drugs. Without identifying themselves, the officers used a battering ram to gain entrance into the young woman's apartment where she was in bed with her boyfriend. Allegedly, the moment after the couple shouted and asked who was there, the police ripped the door off its hinges. Taylor's boyfriend, who had a gun, fired one shot at the police, whom he believed were intruders. The officers reciprocated with several shots fired into the room. Five of the bullets struck Taylor. Three months after the killing, the Louisville Metro Council unanimously passed Breonna's Law banning no-knock warrants. Still, many more states and cities continue to allow no-knock laws to stand with tragic results, a legacy that lives on from July 1969.

10

GROUND CONTROL TO MAJOR TOM

Only days before the launch of Apollo 11, one of the most iconic rock songs of all time was released. Even if you didn't like the rock music of the day, you knew the opening line of this song. Even today, over fifty years later, kids can be heard singing, "Ground control to Major Tom." It was the classic David Bowie recording "Space Oddity."

You can forgive people for thinking the song was about the first moon landing, since it was released only five days before the launch of Apollo 11 on July 11. And in the lyrics, there is one line that sort of gives a nod to the lunar explorers when Major Tom sings, "Here as I'm floating in my tin can, far above the moon." But that wasn't the case. The song begins with a surreal 1969 countdown and launch before Major Tom reports all is well and eventually that he is stepping out of the door of his capsule, presumably for a spacewalk. That's where things take a bit of a dark turn as ground control loses contact with the capsule and frantically tries to regain communications.

In a 2003 interview with *Performing Songwriter* magazine, Bowie set the record straight, saying that the song was in fact inspired by the 1968 Stanley Kubrick movie, *2001: A Space Odyssey* which was based on the popular novel written by science fiction writer Arthur C. Clarke.

"I was out of my gourd anyway," Bowie told the magazine. "I was very stoned when I went to see it [the movie] several times and it was really a revelation to me. It got the song flowing."[1]

Nevertheless, Phillips Records saw a marketing opportunity and released the song just prior to the Apollo launch. Despite a strong marketing campaign complete with video, the song only peaked at number five on the UK charts and barely made a ripple on US charts coming in at number 124.

As a sidenote, when the single was rereleased in 1973, "Space Oddity" reached number 15 in America. When it was rereleased one more time in 1975 by RCA Records, it became Bowie's first number 1 record.

The strong headwind "Space Oddity" faced in becoming a huge hit early on, at least in the United Kingdom, was due in part to the British Broadcasting Corporation, the BBC. At the time, the BBC was virtually in complete control over what the Brits heard and saw on radio and television. When they heard the single, the company immediately banned it from the airwaves until after the flight of Apollo 11, saying that it was in "poor taste" to play it during the mission.[2]

As the saying goes, someone didn't get the memo about the ban and that someone decided the song would make the perfect theme music for BBC TV's coverage of Apollo 11.

"It was picked up by British television and used as the background music for the landing itself in Britain," Bowie later recalled. "Though I'm sure they really weren't listening to the lyric at all. It wasn't a pleasant thing to juxtapose against a moon landing. Of course I was overjoyed they did. Obviously some BBC official said, 'That space song, Major Tom . . . blah, blah, blah. That'll be great.' Nobody had the heart to tell the producer, 'Um, but he gets stranded in space, sir.'"[3]

Decades later, Major Tom really did make it into space when astronaut Chris Hadfield performed the song and filmed an accompanying video during his stay aboard the International Space Station (ISS),[4] making him a YouTube sensation. The haunting video opens with ISS coming out of Earth's shadow before we see the beautiful cloud-filled blue planet far below. From there, the video transitions to Hadfield floating from node to node in ISS singing the song. The astronaut was taken by surprise when he heard that one of his new devoted fans, Bowie himself, described it as "the most poignant version of the song ever done."[5]

It was a baffling year for music with the sugary sweet sounds of the Archies singing "Sugar, Sugar" and the Fifth Dimension singing what became a hippie anthem, "Aquarius / Let the Sunshine In," while at the same time, pop music was also taking a hard turn toward the edgier side with protest songs like Creedence Clearwater Revival's "Fortunate Son" and Edwin Starr's "War!" hitting the *Billboard* Top 100.

Of course, when it comes to music, every decade has had its one-hit wonders and the 1960s, and 1969 in particular, was no exception. One of those popped up on the music charts unexpectedly in July 1969 and took everyone, including its writers, by surprise. It was the number 1 song written by the team of Denny Zager and Rick Evans, better known simply as Zager and Evans. It was titled "In the Year 2525."

"In the Year 2525" is a rather foreboding apocalyptic tale of what the future would be like for mankind with a warning about humanity's increasing reliance on technology taking us from the year 2525 all the way up to the year ten thousand, all with a catchy musical base.

The song was reportedly written in the backseat of a Volkswagen van by Evans in only ten minutes after a long night of partying and marijuana. Evans performed the song with a few bands but never thought it sounded quite right. When Zager heard it, he was intrigued by the lyrics and knew exactly what the song needed—new music.

After revamping the music, the duo played the song live in front of an audience, and according to Zager, "We knew it was special because the crowd looked stunned and wanted to hear it again and again."[6]

While many critics panned the song as being "artless," *Time* magazine did a cover story on the Nebraska songwriters where they first speculated that the song was written by a computer. Then, after recognizing the songwriters' talents, they added the cover headline, "Even the Beatles Would Be Jealous!"

The song was released on June 21 and quickly ran up the charts, hitting number 1 on July 12, where it stayed for six weeks. "No one expects something like that," Zager said in an interview with *Forbes* magazine. "Who would have believed two farm boys from Nebraska would have the number one hit in

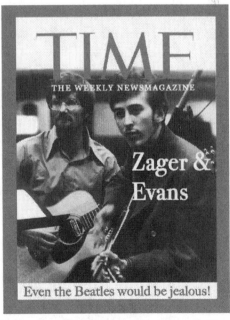

The cover of a July 1969 edition of *Time* magazine proclaims that the one-hit songwriting ability of Zager and Evans would make even the Beatles jealous. *Author's collection*

the world and *Time* magazine saying the Beatles would be jealous. I couldn't have dreamed it."[7]

Sadly for Zager and Evans, the single was their one and only hit, but it was a spectacular ride.

The television landscape in both the United States and United Kingdom in July 1969 was bleak. It was back during a time when the television networks would rely on reruns of a show that had previously aired in the fall of the previous year to pad out their nightly schedule throughout the summer instead of airing brand new shows. The feeling was that most people would be outside enjoying summertime fun instead of watching TV.

On July 12, a television phenomenon and legend was about to be born. A science fiction drama set in the twenty-third century that would cause many of NASA's future astronauts and engineers to dream of going to the stars. The show's story begins three years earlier.

On September 8, 1966, a new series would appear on the NBC television network that its creator, Gene Roddenberry, promised would be a "Wagon Train to the stars," hinting that the show would feature weekly adventures to new and exciting worlds across the universe, much like the old 1950s TV show, *Wagon Train*, which took viewers back to the old west. The show would be called *Star Trek*.

Star Trek brought to television screens across America a look into the future that was both hopeful and promising, while beneath its action-packed façade it subliminally made Americans come face to face with real-life issues that were facing the country during the 1960s, including racism, war, and the Cold War.

Ratings were sluggish, to say the least, and even though the opening credits promised viewers that they would join the crew of the USS *Enterprise* on a five-year mission, that mission was cut short when *Star Trek*'s production company, Paramount, pulled the plug abruptly after only three years on June 3, 1969.

On the other side of the ocean in Britain, another science fiction series was running a parallel course—*Dr. Who*. The show told of the adventures of an eccentric time-traveling scientist who, with various friends tagging along, battled evil robots, monsters, and Time Lords. The doctor could travel back and forth through time in his *TARDIS* (Time and Relative Dimensions in Space) that on the outside resembled a small blue British phone box but inside, the interior was enormous.

Much like *Star Trek*, in July 1969, *Dr. Who* was taking a ratings nosedive, so the BBC decided to put the show on an eighteen-month hiatus telling its producers that when—and if—it returned to the air, its length would be cut in half and the star, Colin Baker, would have to be replaced. Producer John

Nathan-Turner had finally had enough and told the BBC that he was moving on to do other projects, to which the broadcasting giant replied that if he left, he would never produce another show for the BBC and that *Dr. Who* would be canceled.

Like US networks, the BBC would air summer reruns, but they didn't have a show to fill the good doctor's time slot. The company reached out across the Atlantic and picked up rights to air *Star Trek*. On July 12, *Star Trek* replaced *Dr. Who*, and the response was incredible. Viewer reaction was overwhelming and British viewers couldn't get enough of the starship *Enterprise* and its crew.

At the same time, back in the states, *Star Trek* went into syndication so any television station in the country could sign up and air reruns of the original series. From that moment on through today, there has never been a day when the original *Star Trek* has not been aired on a television station somewhere in the United States or the world.

The phenomenon and cult following of *Star Trek* continued to grow exponentially over the years, and today, over fifty years later, it seems that the thirst for more and more *Star Trek* cannot be quenched. Demand for *Trek* has grown so much that there have been thirteen full-length motion pictures made and twelve television series produced including three animated cartoons.

Meanwhile, the music scene in the summer of 1969 was about to get a jolt with a historic music festival that would take place in a small, sleepy village nestled in the rolling mountains of the Hudson Valley of New York State. While the music festival was being billed as "three days of peace and music" and would be called Woodstock, it actually didn't take place in the town of that name. Instead, it would take place on a small dairy farm in Bethel, New York. More on how the festival made it to Bethel is coming up in a later chapter. Right now, though, in this week in July 1969, the town of Woodstock was feeling the burden of being an artistic mecca.

The mountains of the Hudson Valley are simply breathtaking with their rolling ridges of thick, green forest and the mighty and wide Hudson River cutting a swath between them. The region has beckoned the creative talents of countless artisans and musicians for centuries. Here in this expansive landscape, three men—Ralph Radcliffe Whitehead, writer Hervey White, and painter Bolton Brown—established the Byrdcliffe Colony in 1902, a sanctuary, as it were, that writers, artists, and musicians could migrate to and immerse themselves in the solitude and beauty of the valley.

Woodstock has been an artistic and cultural hub ever since, drawing in such celebrity artists as Bob Dylan, dancer Isadora Duncan, and naturalist John

Burroughs,[8] but the townspeople who lived there year-round have always held a weary opinion of those artistic types.

When Byrdcliffe first opened its doors, and artists began moving in, the town would shut down its public swimming pools to keep out the "dirty newcomers." In July 1969, the town was seeing a new migration of artists and groups of teenagers with long hair, beards, beads, and sandals. As it was seventy years prior, the town closed their pools, this time to keep out the "unsavory types."[9]

The town's residents were concerned that these "unsavory types" would bring serious problems to the community. At a town meeting, one resident demanded that the police tighten their control on the youngsters, adding that any of the teens arrested should be "deloused and have their heads shaved to clean them up."[10]

The migration of the hippies from places like New York City's Greenwich Village started only as a weekend occurrence. Now these youngsters were making the decision to stay. In the center of town, a large grassy area known as the Village Green that was surrounded by quaint shops and Dutch-style houses and a church suddenly had new shops springing up where the sound of loud electric guitars and thumping drums vibrated the town.

The police began ramping up arrests as they kept a close watch on the newcomers. If any of them were seen loitering or trespassing, they were immediately put behind bars.

But why were these youngsters flocking to Woodstock? As a fifteen-year-old African American boy from New York City's Lower East Side told a *New York Times* reporter, "It's beautiful man. I mean the place is beautiful. The people are beautiful."

And those teens were not as destitute as many of the townspeople believed. Many of the "migrants" were buying $30 to $60K homes in the surrounding mountains. As one town lawyer commented, "they have more money than us."[11]

The onslaught of young hippies to Woodstock and the Hudson Valley would continue throughout the summer and into the fall of 1969. As the deluge of young people invading the town grew, a group of entrepreneurs went to the city council to propose a three-day "peace and music festival" be held in Woodstock, the likes of which had never been seen before. City officials quickly shut the proposal down.

The trouble was that the organizers had already been selling tickets. Big-name artists were already signing up to be a part of the event. They had the bands but didn't have a place to stage the concert. They decided to try and gain the support of the nearby town of Wallkill, but those townspeople would also have something to say about that themselves in the weeks to come.

A FOOTNOTE TO HISTORY

July 8: A highly combustible and highly controversial weapon had been deployed in the Vietnam War—napalm. Napalm was a chemical combination of gasoline, benzene, and polystyrene that formed a gel, sort of like a jelly gasoline, that when dropped from the air, would turn into liquid fire and burn large swaths of foliage. Sadly, it would also stick to and burn humans to the point where they looked like "swollen, raw meat." Far from Vietnam, Long Beach, California, narrowly escaped the detonation of napalm as four canisters of the material rolled off a truck that was carrying them as it was entering a freeway. Fortunately, none of the canisters detonated.[12]

July 9: You would think that since French is one of the two predominant languages in Canada, the other being English, that it would be recognized throughout the country's federal government system, but that was not the case. In October 1968, the Official Languages Act was introduced to Canada's Parliament that would require French to be recognized as a coexisting language with English. This meant that debates and documents written in French would be just as legal and recognized as English versions.[13] The act was passed on this date in 1969.

July 9: Hundreds of hungry rats were driven out of Military Park in Newark, New Jersey, by police officers. According to the AP report, the policemen used shotguns to drive the hoard out, killing twenty-three in the process. Responding police said that they had to be selective in their targets for fear of shotgun pellets bouncing off sidewalks and wounding passersby.[14]

JULY 13–19, 1969

Liftoff! We Have Liftoff!

11

OUT-OF-TOWNERS WHO CLUNG TOGETHER

As was the case in many cities across the South, Huntsville, Alabama, was just a sleepy little farming town whose economy was based on either King Cotton or the harvesting of watercress. The European and Asian watercress with its peppery flavor had become increasingly popular in a variety of American dishes, and the plant found a particularly favorable home in the Huntsville area. The myriad of creeks, streams, and rivers were perfect for growing the leafy vegetable, and the town became known as the Watercress Capital of the World.

That all began to change in 1941 when the US Army combined two of their facilities in Huntsville to create the Redstone Arsenal, or as many of the old-timers called it, the "bullet factory." At first, Redstone was assigned to making ammunition and toxic chemicals for the military, but in 1948, its mission changed to making missiles, and it became part of the Army Ballistic Missile Agency (ABMA). It was here that a group of captured German rocket scientists (designated as "prisoners of peace") who had developed the V-2 rocket for Hitler during World War II arrived to assist the US Army in their own missile program. Dr. Wernher von Braun and his team would develop a short-range ballistic missile, the Redstone rocket, that would later be transformed to put America's first astronaut into space and eventually become the basis for the Saturn V rocket to put men on the moon.

Huntsville put watercress and cotton in the rearview mirror and never looked back as it transformed into Rocket City. A young Jim Felder and his family were living in Mobile, Alabama, as Huntsville was making the transformation.

To say that Jim's dad, James Thomas Felder Jr., was into electronics is putting it mildly. While in high school in 1940 in Birmingham, Alabama, James and a friend designed a complicated method of communicating with each other between houses by using microphones and incandescent lightbulbs. It didn't work, but the basic concept was proven correct.

During World War II, James was stationed in Mobile at Brookley Air Force Base. The elder Felder was a sergeant in charge of communications during the war for the 1343rd Engineering Battalion where his company commander would tell the troops, "Wherever we stop, you put a phone in my hand that will somehow connect me to anybody I want to talk to."

Following the war and back in civilian life, there were few job opportunities in Mobile until the Navy contracted a company to extend telephone communications to the Navy's test grounds on Santa Rosa Island, Florida. They did so by using microwaves and creating the first wireless telephone system over water.

It wasn't long before Jim's father heard that Wernher von Braun and ABMA wanted to open a computer department in Huntsville to handle telemetry received from their early rockets. Jim's dad loaded the family into the car and made the move to Huntsville, and the rest, as they say, is history.

"One day [in 1959] my father went to work at ABMA," Jim said, "and when he came home, he had a new badge—it said 'NASA.'"[1]

From that moment on, James Felder worked with the NASA team in Huntsville to help develop the digitalizing of that data, making it more useable in a timely fashion. "He would tell us about working on rolls of data on computer paper that he called 'toilet paper' that they would roll down the hall to work with. There was an atmosphere of pride in the work they were doing and a feeling that there was no problem that couldn't be solved."

According to Jim, life in Huntsville in 1969 was like that in any other city, with one caveat. "[The kids] went to dances, thought about keeping their grades up and going to college," he said. "We would all go and have something to eat at Boot's, maybe an ice cream or Dip Dogs at Zesto's, but everyone was somehow connected. Not everybody's parents worked [in the program], but everybody knew somebody who worked there."[2]

Even if your family wasn't involved in the space program, Jim says that the city became immersed in it. "Everyone was into rocket clubs and amateur radio," he said. "Every classroom had a transistor radio in it that was tuned in to every launch and mission. Everything would stop."

Jim would have the chance to meet Wernher von Braun one day during a Boy Scout trip to tour one of NASA's rocket engine test stands on the base. "I remember von Braun talking to the group, and then he took out a piece of blue paper and handed it to me. It said, 'Going to Mars in your lifetime.'"[3]

Jim Felder graduated high school in 1968 and attended the University of Alabama, Huntsville, where he worked on an arts degree. He clearly remembers the day that Apollo 11 landed on the moon. "I was with my girlfriend, Donna, who later became my wife. We're still married to this day. [That night] we settled in on the family couch where the television sat on an expansive stone hearth. Donna's mother had bought an assortment of doughnuts from Winn Dixie for the occasion. We munched on the doughnuts and watched the fuzzy, but ever-so-impressive, video of the landing. Donna left my side to go out on the driveway to look at the moon. It drove home the concept that people we knew, our parents included, had a hand in what was happening."

Not long after college, Jim and a friend opened a business where he worked in marketing communications. That's when his experience with the Rocket City and von Braun came full circle. In the 1950s, the Walt Disney Company arranged a deal with von Braun to create a series of television shows that would outline the rocket scientist's dreams for our future in space. Von Braun was well known for his scientific abilities, his personal charm, and his uncanny public relations skills. But von Braun was much more than that. He was also an artist who could visually bring his ideas to life, and for the Disney specials, he would create large scale models of his vision, including futuristic rockets and a circular space station that resembled a wheel turned on its side, a concept that was later used in the production of the movie *2001: A Space Odyssey*. Von Braun envisioned the space station continually rotating in space. The centrifugal force produced by the spinning would create artificial gravity for astronauts on the space station.

In 1994, IBM approached Felder and asked his company to help restore the model of von Braun's space station for an upcoming exhibit. With painstaking care and a lot of BASF film cleaner, Felder and his team repaired the model, which made it to the Smithsonian museum in Washington.

The towns of Titusville and Indian River City in Florida could not have been more different than Huntsville. When Shirley Wildermuth and her husband, Wayne, arrived on New Year's Day in 1966, there was little there.

"It was a culture shock," Shirley recalled in a NASA oral-history interview. "There were very few stores. There was the Big Apple Shopping center, two or three bars, and a couple of restaurants. [When we went to eat] we would go to Cocoa Beach and eat at Ramon's or the Mouse Trap. Ninety percent of the people in town were associated with the space program," she continued. "There were very few locals. They were all 'newcomers,' 'out-of-towners' who clung together."[4]

Shirley and Wayne were among the hundreds of contractors brought in to aid in America's space effort. While working with the Bendix Corporation in San Diego, Wayne helped develop a clean room where parts could be micro-cleaned, something that would be extremely important to the space program. He was spotted by a recruiter, and the next thing the couple knew, they were moving to Titusville to work as NASA contractors.

Shirley assumed she would be just a stay-at-home mom but soon found that there wasn't much to do in Titusville, so she applied for a job with Bendix in NASA's Manned Spacecraft Operations office. She was hired on as a component specialist where she ordered and dispersed parts. Shirley recalls the difficulty she had in getting a raise. "[When I asked about a raise] the supervisor told me, 'You can't get a raise because you're female, and females can't order parts.' I looked at him and said, 'What do you think I'm doing?'"

Later when Bendix lost the contract, she hired on with Boeing and began work in their warehouse at Cape Kennedy, where she and her friend Joyce Travis became the first women warehouse people (not warehousemen) at the Cape. They would stock shelves, sweep floors, drive trucks, and deliver parts. "I remember that on Secretaries Day, the company offered us flowers," she recalls. "It didn't go over very well. We had to fight all along the way to be equals."[5]

The atmosphere in the towns around the Cape in 1969 was electric. "It was thrilling coming to work every day knowing that you are part of the space program. Everybody out there had one objective in common. It was a large extended family that lived, breathed, and talked space."

The most memorable moment of Shirley's career came on July 16 with the launch of Apollo 11. "We would go to the roof of the MSO building and could look straight down to the pad. The launch was absolutely superb. There were tears, goosebumps, choked throats. Then the rumble of the noise and the vibrations. It is something that you've got to experience to really know how tremendous it was. I still get goosebumps knowing we had done this."

One of the most difficult aspects of the Apollo missions to simulate was the actual landings on the moon. The lunar module was designed to only fly in the one-sixth gravity of the moon, not on Earth. So NASA developed what was affectionately called the "Flying Bedstead," a spindly, spider-like contraption with a series of rocket engines that would allow astronauts to train on Earth to land the lunar module. It was actually called the Lunar Landing Research Vehicle (LLRV).

It was an unwieldy beast that nearly claimed the life of the first man on the moon when it went out of control during a training session at NASA's Flight Research Center at Edwards Air Force Base in Kern County, California. The

While women still had a long way to go, in 1969, they were steadily making inroads into technical careers across the aerospace industry both at NASA and with contractors for the Apollo missions. Pictured is test engineer Monica Koerner from a Bendix Corporation booklet outlining the Apollo 11 mission and the role the company played in it. *Bendix Corporation*

one non-astronaut who had the privilege—if you can call riding a runaway roller coaster a privilege—was Scott McCleod, who had fond memories of life near the Air Force base as well as at the other NASA centers across the country, as he worked with engineers and the astronauts to develop computer simulations of the lunar module and flying it.

McCleod was a Korean War aviator who once won a special commendation for downing five planes. It was a cardboard award made by his fellow pilots for his crashing five US planes. After the war, a friend of his at the time who worked at the predecessor to NASA, the National Advisory Committee for Aeronautics (NACA), helped him land a job with the agency as a test pilot. From there, the sky was literally the limit for Scott.

"[The LLRV] was like flying a helicopter," McCleod said. "It was very similar. I flew many simulated flights to the moon, even on T.V."[6]

What McCleod was referring to was his appearances during CBS News coverage of the Apollo 11 landing. The network had set up a mockup of the lunar module in a studio, and since there was no actual live television of the landing itself, the network would cut to McCleod in the mockup acting out what the astronauts were doing.

McCleod met his wife, Joyce, in 1968. One day, McCleod and a team from NASA were looking into the future and were discussing how to get to Mars and what the crew makeup would be. It would be an eleven-month journey that would pose many challenges for the astronauts.

"[If there was] one [astronaut], he would go crazy. Two, they would argue all the time. Three and it would be two against one. Someone asked, 'Would it be all men?'"

Realizing that women would also be on such a mission, they decided that they should visit a woman they had heard of who had experience doing hazardous, long-duration, deep-water research in a submarine.

"I went on a search to find her," McCleod says with a grin. "I walk in and see a pretty girl sitting at a typewriter. Probably typing a letter to her boyfriend. I asked if I could see the submarine. She said, 'I'll show you when I'm done with this.' I said, 'Isn't there a man who could do that?' She said, 'No! I'll do that when I finish this!' We fell in love right away."[7]

Flying was in Scott's blood, and every now and then he had a special guest join him for a ride. One day, one of NASA's directors, James Fletcher, asked McCleod to take him for a ride. They climbed into an SNJ trainer plane and took off for a half-hour joy ride full of aerobatics. When the ride was over, McCleod lowered the landing gear to land, but one of the two wheels did not drop down.

Thinking quickly, he hit the good wheel hard on the runway thinking that it would dislodge the other, but it didn't. He would have to dig one wing into the dirt next to the runway for an emergency landing.

Little did he know that CBS news anchor Walter Cronkite was at the airport filming segments for the nightly news. Cronkite turns the camera at the incoming plane and says, "Scott McCleod . . . [will] be trying to make a landing now. Stay tuned. We'll catch him as he digs his wing tip in and rolls up in a big ball of flame."

McCleod skillfully landed the plane as he had planned, digging one wingtip into the dirt causing him to spin around and come to a stop. He spent the rest of the day late into the night being debriefed about the flight when a man from the control tower told him about Cronkite. His crash would be on the news and hopefully his wife hadn't seen it. He called his wife, who was nine months pregnant, to tell her he would be late and asked if she had seen the news. She hadn't.

He finally arrived home and Joyce prepared him a martini. She stood next to him and asked how his day was. "I said it was fine. Then she asked, 'anything happen today?' I said, 'No. Nothing happened.' She said, 'are you sure?' And I told her again that nothing happened. 'I was watching the news tonight!' and she poured the martini over my head."[8]

Unlike the rural towns of Titusville and Huntsville, Houston, Texas, was already a booming city that was built on the burgeoning oil industry. It didn't take long before the city became known as the energy capital of the world, but in 1965, a new high-tech industry had moved in that helped bring a new industry to the city—the opening of the brand-new NASA Mission Control Center (MCC). Suddenly, a new breed of workers was flocking to the city—engineers and astronauts.

The provost and vice president for academic affairs at Framingham State University in Massachusetts, Linda Vaden-Goad, grew up in the city during the Apollo missions.

"I was sixteen years old and in high school," Goad said. "I was a serious tennis player. It was exciting to be in Houston with all of the talk around space and a lot of them [astronauts] lived in Houston."[9]

Linda remembers sitting around the living room with her family as Apollo 11's *Eagle* landed on the moon. "We watched everything about it, everything we could see. The first word out of Armstrong's mouth during the landing was 'Houston,' so for those of us growing up there, that was an added piece of the whole story. It gave us a sense of identity. These were our neighbors who were going into space. Their families lived around us. It was an exciting time. My childhood was defined by it."[10]

Even though Linda and her classmates were not part of the space program, and unlike in a small town like Huntsville where everybody knew everybody, there wasn't a direct connection with the space program, but the sense of accomplishment that her city brought to the world was overwhelming.

"To me, growing up in Houston and being part of that excitement of seeing new worlds, the fact that people could do things that in the past you couldn't imagine. It made the moon real. The fact that they did that, that they did that in my formative years, it gave us the feeling that we could do almost anything."

12

T MINUS THREE DAYS

Launch day was drawing near. July 16 would see the incredible machine that would take humans on their first otherworldly trip come to life on the beaches of Cape Canaveral, Florida. America's competitor in the space race, the Soviet Union, had lost the race to put the first man on the moon, but in a last-ditch effort to at least be the first to bring a piece of the moon back home to Earth, the Russians launched Luna 15 three days before the Apollo 11 launch.

Luna 15 would not be the spectacle of landing humans on the moon, but it would be a historic event, nonetheless, by doing something no other spacecraft had done—landing on another celestial body and returning samples to Earth. In that sense, the unmanned spacecraft's mission would beat the Americans.

In theory, Luna 15 would fly to the moon, land on its surface, scoop up some rock and soil samples, then return them to Earth days before the Apollo 11 astronauts would return. After its successful launch, the spacecraft was put into a temporary Earth orbit before its engines fired, sending it sailing toward the moon.

Back in the United States in the burgeoning town of Huntsville, the residents were getting ready to celebrate the launch of Apollo 11, but the excitement they were experiencing was quite different from that of the millions of people who were flowing into Brevard County, Florida, to witness the launch from Florida's sunny shores.

Nestled in the windswept valleys formed by the rolling mountains of the region, Huntsville's Marshall Spaceflight Center had brought thousands of contractors together from around the country to design the mighty Saturn V rocket that stood poised on the edge of the Atlantic Ocean to open a new and historic chapter of human exploration.

The two hundred thousand men and women who had helped design and build the rocket were noticeably on edge the morning of July 16, as a second hydrogen leak was detected in the rocket while it was being fueled some seven hundred miles away. As it turned out, the leak was exactly the same as the one that had occurred during the wet dress rehearsal of the rocket the prior week. Engineers at Cape Kennedy had gleaned valuable information from that experience and the leak was quickly resolved.[1] With the leak repaired, the atmosphere at Marshall became much lighter, and the team's otherwise muted voices spoke out confidently about the upcoming launch.[2]

The rocket was the brainchild of German rocket scientist Dr. Wernher von Braun, who was brought to the United States with his fellow scientists and engineers to work on the country's military rocket program. Von Braun always had his sights set on the stars and sending men to the moon and beyond. Along with his fellow rocket scientists, von Braun had created a series of reliable rockets for the United States, complete with state-of-the-art guidance systems that would be the building blocks for the Saturn V, which was now standing on launch pad 39A. Those early rockets developed by von Braun, however, had been turned into a new and terrifying weapon of war by the German government—the Vengeance Weapon 2, better known as the V2 that rained down on England during the war, killing nine thousand civilians, not to mention the twelve thousand enslaved laborers that died during production of the rockets. Once brought to the United States, von Braun's mission quickly changed from America's military rocket program to one of beating the Russians to the moon after the Soviets had launched the world's first artificial satellite, Sputnik 1, in 1957. It was von Braun's team who gave the US response when his Jupiter rocket launched America's first satellite, Explorer 1, in 1958.

With his well-known confidence and swagger, von Braun told reporters that, yes, there was always the chance that something could go wrong with the launch, but he had no doubt it wouldn't.

Besides the throng of people who had invaded Titusville and Cocoa Beach for the launch, the US Navy reported that eight ships of the Soviet fleet were steaming to a location twenty-five miles southeast of Miami, Florida, a location that would offer them excellent viewing opportunities of the launch. The US aircraft carrier escort, the USS *Gary*, was dispatched to shadow the fleet and keep an eye on their movements during the launch attempt.[3]

In Houston, Texas, at the Manned Spaceflight Center, lunar module manager Carroll Bolender had nagging feelings about the lunar module (LM) that was tucked away inside the Saturn V that would—if all went well—land Neil Armstrong and Buzz Aldrin on the moon on July 20.

The two lunar modules used during test flights on Apollo 9 and Apollo 10 and the actual spacecraft that would land the first humans onto the moon had initial quality issues that could jeopardize the landing. For example, Bolender had found that much of the wiring on several of the vehicles was broken.

Grumman, the company that manufactured the lunar modules, made necessary repairs, and during the final flight readiness reviews prior to the launch of Apollo 11, Bolender and flight safety director Martin Raines reported that the general quality of repairs on the landers had improved greatly but that there were still twenty-three issues that needed to be ironed out.[4]

Then there was another concern that the associate administrator for manned spaceflight, George Mueller, brought to the attention of management. He had concerns that the flights following Apollo 11 were coming in such quick succession that fatigue might overwhelm the workers back here on Earth,[5] posing a possible safety risk.

In response, director Robert Gilruth said, "Worry [along with me] but don't allow [it] to interfere with driving your staff at full throttle until . . . the lunar landing."[6]

And with that, the readiness review teams came to a unanimous conclusion: the only great risk they could see in the flight of Apollo 11 was that they would make the first-ever lunar landing, and that was a risk that would be there no matter what equipment was used or what crew made the voyage. Apollo 11 was given the go-ahead for launch at 9:32 a.m.

As the countdown ticked on, the tension continued to mount, and at that moment, everyone from the engineers in the firing room to mission control in Houston to the average person on the street realized just how dangerous a spaceflight is from launch to splashdown. It is so dangerous that the Apollo 11 astronauts, and all of the astronauts in the early space program for that matter, found it impossible to obtain life insurance. To care for their families in the event of a disaster, the astronauts created their own form of insurance—they would autograph a series of envelopes known as "covers."

Covers are cached envelopes with a unique image, stamp, or both on them that relates to an event in history, anniversary, or person. Most of the time, the envelopes used by astronauts carried their mission logo on them. What makes the envelopes special is the postmark. The astronauts would autograph the envelopes then have them postmarked at the Kennedy Space Center's post office the morning of their launch. The envelopes would then be stashed away in a safe-deposit box, and if an accident occurred and the crew perished, the envelope would become highly valuable and fetch an incredible price at auction, which would sustain the families the astronauts left behind.[7]

In Firing Room 1 at Cape Kennedy, the tension was mounting as the countdown clock reached T minus 22 minutes. "It's in the last 22 minutes that the countdown really gets dynamic," launch director, Rocco Petrone told reporters. "That's when high pressure hydrogen starts flowing into the tanks, and from then on, the guys are watching the temperature gauges and everything like hawks. Those last 22 minutes are when we really earn our pay. I'd have to say," Petrone continued, "that you could feel the tension. The people knew this was the big one. There is a certain amount of, shall we say, static electricity in the air."[8]

Prior to climbing aboard the capsule, command module pilot Michael Collins commented on the crowd that waited with supercharged excitement along the space port's beaches. "It seems like the emphasis is in the wrong place somehow," he was quoted as saying. "Like maybe we ought to wait until the flight's all over and see if this grand and glorious thing works out."[9]

The clock continued to silently count down, with only a few built-in holds pausing the count so that the rocket's systems could be given the once-over by engineers to make sure all was ready before the countdown picked up again where it left off.

Inside Firing Room 1, a sea of men wearing white button-down shirts and dark ties checked their consoles regularly for any anomalies that could spell disaster for the launch. One image stood out in the sea of men in the control room. It was one lone woman—JoAnn Morgan.

JoAnn Morgan was born in Alabama but moved to Titusville, Florida, in the middle of her junior year of high school. From her new school, JoAnn could watch launch after launch from the Cape. After the launch of America's first satellite, Explorer I, and the discovery of the Van Allen radiation belts that circled the Earth and protected all life on it, she thought, "This is profound knowledge that concerns everyone on our planet and I want to be a part of this team."[10]

Her chance came when she spotted an advertisement that the Army Ballistic Missile Agency (ABMA) had placed in the local paper looking to recruit two students to work as aides to its engineers.

"Thank God it said 'students' and not 'boys,'" she later said, "otherwise I wouldn't have applied."

Not long after hiring on, ABMA was rolled into a new agency specifically designed for the exploration of space and aeronautics—NASA. During this time, Morgan worked hard to receive a Bachelor of Arts degree in science. Men like Dr. Wernher von Braun and Dr. Kurt Debus saw the young woman's

JoAnn Morgan was the only woman in the launch firing room during the launch of Apollo 11. *NASA*

drive and determination and in particular her experience with writing technical papers, working with data systems, and building computer components. With their guidance, the men helped her build upon her experience and education until finally she was promoted to senior engineer.

Unfortunately, this was the 1960s, a time when women were held back by men from achieving their full potential. Stereotypes and sexism ran rampant, but JoAnn had determination as well as strong backing from her managers. Early in her career, her manager, Jim White, called a meeting of launch engineers. She was excluded from the meeting but for a good reason. White addressed the men in the room.

"This is a young lady who wants to be an engineer," he said. "You're to treat her like an engineer."

One of the men in the room chimed in. "Can we ask her to make coffee?"

"No!" White shot back. "You don't ask an engineer to make coffee!"

As the launch of Apollo 11 drew near, unbeknownst to Morgan, the man who developed the launch processing systems for Apollo, Karl Sendler, had a meeting with the Kennedy Space Center's first director, Dr. Kurt Debus. When the meeting was over, Debus met with Morgan.

"You are our best communicator," he told her. "You're going to be on the console for Apollo 11."[11]

And with that, JoAnn Morgan made history by becoming the first woman in the firing room for an American rocket launch.

In the Lakeland, Texas, suburb near Houston where most of the astronauts and their families lived, the streets became eerily silent and deserted. The normal sound of children running, playing tag, or riding their bicycles vanished, leaving the community looking like a ghost town. Everyone had moved indoors to watch the spectacle that was about to unfold.[12]

With nine minutes left, the engineers in Firing Room 1 were polled. It was a rapid fire session where the Spacecraft Test Conductor (STC) calls out each of the team's stations expecting an answer of "go" for launch or "no go."

STC: "Booster?"

Booster: "Go!"

STC: "FTS?"

FTS: "Go!"

STC: "CFC?"

CFC: "Uh, stand by STC. We don't have VHF or S-Band uplink going. We're checking."

It turns out the communications error was minor.

STC: "Verify go for launch?"

Launch Director: "Go for launch."

NASA public affairs officer Jack King, who has been updating the gathered crowd at the Cape and the millions watching on television with a minute-by-minute description of what was happening began his final announcements, "T minus 15 seconds. Guidance is internal."

King continues. "Twelve, eleven, ten, nine, ignition sequence starts . . . six . . ."

At this point, the five enormous Saturn V engines burst to life, belching out a thick cloud of black smoke before a brilliant orange fire ball engulfs the bottom of the rocket.

The commentary continues without hesitation, ". . . five, four, three, two, one, zero."

Thousands of VIPs watch the launch of Apollo 11 from a location only 3.5 miles away from launch pad 39A. *NASA*

At this point, King, whose coolness under pressure has been described as being "legendary,"[13] gets a bit choked up as he announces in the singular, "all engine running" instead of "all engines running."

Massive steel clamps that tether the rocket to the launch pad swing back and the towering stack of machinery begins a slow ascent.

"Liftoff! We have liftoff! Thirty-two minutes past the hour!"

NASA public affairs officer Gene Marianetti perfectly described the feeling as he watched the launch.

> You see the ignition of the main engines, and then it seems like it just sits there. There's anticipation. It's building up thrust—7½ million pounds of thrust. Then the clamps release. It lifts off so slowly. You get the sound like a deafening shock wave when it's already 400 feet up. It's a beautiful sight, hard to describe. Twelve million parts and everything worked perfectly.[14]

Apollo 11 was on its way, building up speed incrementally that would eventually put the astronauts in an orbit around the Earth before they would fire the third stage engine and head for the moon.

Next to riding atop what could ostensibly be called a giant stick of dynamite, the most dangerous part of the launch is getting the rocket past the launch tower. The rocket could sway or veer slightly and collide with the massive steel structure. There is a palpable sigh of relief when the launch team and Jack King finally announce, "Tower cleared!"

After spending just over two and a half hours in Earth's orbit, the Saturn's third-stage engine was ignited, sending the astronauts off on their journey to the moon. For the next two days, the crew of Apollo 11 would perform routine housekeeping to keep the capsule on target and verify that all systems were ready for the big day—the first lunar landing by humans. They also turned on their capsule's television cameras, giving the world a front-row seat to life aboard *Columbia* and images of Earth, the "pale blue dot," as planetary scientist Dr. Carl Sagan would call it years later, hanging in the inky blackness of space.

As *Columbia* and *Eagle* zoomed serenely through space toward the moon, the Soviet Union's Luna 15 had arrived in lunar orbit. Upon arrival, Soviet scientists looked at photos taken by the spacecraft of the lunar surface and determined that the lander's proposed landing site was too mountainous and therefore too dangerous to attempt a landing. Instead, they opted to take their time and orbit a few days longer before deciding where they would land the craft and bring the first lunar rock samples back to Earth before Apollo 11.[15] If they could pull off the mission, it would be a far cry from actually landing a man on the moon, but it would still be a remarkable feat and would steal headlines away from Apollo 11, albeit for a fleeting moment.

Three days after the launch, the weekly edition of *TV Guide* magazine hit newsstands, its cover emblazoned with its traditional red and white logo and the headline, "The First Live Telecast from the Moon." Below the headline there was a full-color artist's rendering of what it would look like when Armstrong and Aldrin took those first steps on the moon.

Inside, the magazine reviewed how each of the television networks would cover the historic event and provided readers with a minute-by-minute description of the astronaut's activities on the lunar surface so that viewers could plan their watch parties.

The magazine's editors appropriately pointed out that television executives tended to throw the word *special* about indiscriminately to describe many of their TV shows. It seemed like every show was a "special episode," but this telecast would be different. *TV Guide* called it the most "special special" ever.[16]

13

WHAT THE HECK
IS GOING ON?

Following the overthrow of the Shah of Iran by Islamic fundamentalists in 1978, the pervading view that swept across the parched desert landscape of that country became one of virulent criticism of the United States. This vitriol led to one of the most dramatic and tense 444 days in US history when a group of Iranian students seized the American embassy there, taking more than fifty Americans hostage. From that moment on, the two countries have been adversaries on the world stage with Iran threatening the world with a proliferation of more nuclear weapons and a complete and total crackdown on human rights within its borders. Women in particular have faced harsh penalties for a number of "violations" of Islamic law, including not wearing a hijab to cover their hair correctly.

In September 2022, a twenty-two-year-old aspiring lawyer, Masha Amini, died in police custody while she was being held for breaking the country's dress code for women. For the women of a country, who had been under the thumb of the fundamentalist government since it first took hold in 1978, Amini's death was the match that lit the fuse. Iranian women had had enough and took to the streets. Thousands of women, later joined by men, marched in protest, burning their hijabs in open defiance of the government. Many were arrested and many died, some through public hangings. With the protests came a crackdown on information coming out of the country. Journalists were expelled from the country. Internet access was cut. The only way the rest of the world knew of the protests was through videos recorded on cell phones that were smuggled out of the country.

For a country looking to turn the spotlight off of the protests and get a little good public relations on the world stage, the protests couldn't have happened at a worse time for Iran. It was time for the FIFA (Federation Internationale de Football Association) World Cup Soccer tournament, an event that is four years in the making, where the last thirty-two teams from countries around the world still standing after three years of competition put it all on the line. Followers of the World Cup are fanatical about it and go all out in support of their country's team. In the 2022 edition of the tournament, the two adversaries—the United States and Iran—would face off against each other for bragging rights.

The result wasn't exactly what the Iranian government was hoping for, as their team lost to the Americans 1–0. Videos smuggled out of Iran began trickling into news outlets around the world. The clips showed not a dejected and beaten Iran, but thousands of people rejoicing—the United States had beaten their home country. Fireworks went sailing into the sky. Crowds chanted "U-S-A." Horns blared in the streets. Columnist Bobby Ghosh reported on the videos to viewers on the cable news channel MSNBC. "That is something I would not in a million years have expected to see in my own lifetime," he said, "which tells you that they hate their regime and they hate the team that represents that regime."[1]

And that's how soccer (or football as it is known elsewhere around the world) is. It is an international sport that spawns nationalistic fervor among fans. The game is cherished by a nation's people to the point of fanaticism, with the fans' unbridled jubilation sometimes spawning deadly riots and even war, as was the case on July 14, 1969, following the three-game World Cup qualifying series played between the Central American countries of Honduras and El Salvador. Well, that's not exactly true. There was more to the brewing war that would play out in July 1969 between the two countries, but the tournament did play a dramatic part.

Tensions between the two countries had been mounting for quite some time prior to the soccer matches. In both countries elite landowners controlled most of the land used for farming, leaving small family farms in the lurch. Honduras, being a much larger country in size, had many more opportunities for smaller farmers to own land and grow fruit for US fruit companies. Throughout the 1960s, El Salvador saw a sizeable migration of its small, local farm population to Honduras, which the El Salvadorian government and wealthy landowners of the country fully supported, since that freed up more land for the elites to acquire for themselves.

By May 1969, over three hundred thousand El Salvadorans had migrated to Honduras. Rumors were running rampant through the cities and villages in

Honduras that these new immigrants were guilty of bringing with them rapists and murderers. The Honduran government did not take kindly to the influx and began a campaign to deport them. Flyers were distributed around the country that read: "Guanaco [a slang word for Salvadoran]: If you believe yourself decent, then have the decency to get out of Honduras. If you are, as the majority are, a thief, a drunkard, a lecher, crook or ruffian, don't stay in Honduras. Get out or expect punishment."[2] To top it off, the two nations were embroiled in a bitter dispute over who controlled a number of islands that straddled both of their coasts in the Gulf of Fonseca.

On June 2, the Honduran government began deporting the first group of immigrants back across the border. Red Cross relief stations positioned inside El Salvador along the border were overwhelmed with refugees. Almost immediately, the small country of El Salvador began feeling the pressure of the rapidly increasing size of its population on its infrastructure and farmland. The president of El Salvador, Fidel Sanchez Hernandez, was feeling pressure himself from the country's wealthy landowners, who were nudging him toward taking military action against their neighbor to protect their interests. Skirmishes along the border flared with rifle and mortar fire.

Six days later, the two countries would meet face-to-face on a soccer field in the Honduran capital of Tegucigalpa for the first of a three-game series that would determine which country would send the first ever Central American team to the World Cup.[3] The home team won the first game in overtime 1–0. Following the game, fans of both sides spilled out into the streets and clashed in a riot.

One week later, the venue switched to El Salvador, where fans of the home team accosted the Honduran team at the airport. Throughout the night, thousands of fans paraded past the hotel where the Hondurans were staying in an effort to create such a racket that the team's sleep would be disrupted so much that they couldn't concentrate on the field the next day.

The following day, El Salvador won the second game 3–0 to tie the series at one game apiece. Again, violent clashes raged out on the streets following the game. In Honduras, angered fans began terrorizing any Salvadorans living in the country to the point where thousands had to flee for their lives back across the border.

The third and final match would be played on July 27 in Mexico City to determine who would advance to the World Cup. El Salvador won the game in overtime, and immediately following the game, both countries severed all diplomatic ties with each other. In reporting the event, United Press International (UPI) gave the match a name. They called it "The Soccer War." But the fuse had already been lit long before the soccer series began.

Dan Hagedorn was enlisted in the US Army at the time and was stationed at a base near the Panama Canal. His office was equipped with a teletype machine that would kick on and begin automatically typing out messages whenever an alert was posted by the US military or government. On July 17, the teletype began "chattering" frantically.

"I remember it vividly," Hagedorn recalled in his book, *The 100 Hour War*. "It started chattering and it kept on going, kept on going. I said, 'What the heck is going on?' So we went over and took a look at it—that's when we realized the Salvadorians had invaded Honduras."

Indeed, they had. El Salvadoran fighter planes laden with bombs made incursions into Honduras, followed by an invasion by the Salvadoran army. In retaliation, Honduras began strategic aerial bombings of power plants and oil refineries in El Salvador.

The battle between the two countries was brutal and continued for four days before both sides finally ran out of ammunition. On July 20, a ceasefire was called, and the battle, which also became known as the 100 Hour War, was over. In the end, over two thousand were killed and three hundred thousand Salvadorans living in Honduras were displaced.[4]

Across the Atlantic Ocean, one of the United Kingdom's territories, Northern Ireland, was experiencing its own bloody conflict, which became known as the Troubles. But to learn about the events that occurred in July 1969 during what is known as Marching Season, one needs a little background.

It all begins several hundred years earlier during a tumultuous struggle for the British crown in what was called the Glorious Revolution of 1688 in which the Protestant national leader, William of Orange, deposed Catholic King James II. James vowed to return and take up the matter again in Ireland, which he did on July 1, 1690, during the Battle of the Boyne. When the smoke had cleared on July 12, James's attempt at regaining his crown was thwarted by William of Orange, and with the victory, the role of Anglican Protestantism was elevated on the island.[5]

In 1795, the Loyal Orange Order, or simply the Orange Order, was founded. The Order was primarily made up of working-class people with a mission to protect Protestantism in Ireland from perceived threats by Catholics. From that moment on, the months between April and July each year have been filled with celebrations and parades known as Marching Season to commemorate the actual date that James II was defeated, July 12 being the highlight. On that date (which is known, appropriately enough, as the Twelfth), thousands take to the streets in a full day of celebration, with bonfires, fireworks, and parades. Young and old members of the Orange Order wearing bright orange sashes march through the streets to the blaring sounds of pipes and drums. But there was more to the celebration than just celebrating an ancient war victory. The Twelfth also served as an irritant to the Catholic population, taunting them with the fact that they were the minority in Northern Ireland.

After centuries of British rule, Ireland began its split into two separate countries in 1919 when an Irish nationalist political party, Sinn Fein, declared the island an Irish republic independent from the monarchy. In the midst of what was called the War of Independence, the British government enacted the Government of Ireland Act, which partitioned the island into two separate entities, each with its own parliament. There would be a northern province whose counties were predominantly Protestant unionists (loyalists) while the southern province was comprised of mostly Catholic nationalists (republicans).

The Irish republicans opposed the proposal and in 1922 declared that the entire island would be the Irish Free State. Those in the northern province immediately rejected the notion of becoming a free state and withdrew from it, officially becoming what we now call Northern Ireland, and accepted a role of self-governance within the United Kingdom. The southern provinces would break free of the United Kingdom and become a free and separate country on its own, the Republic of Ireland.

As you can see, religion has always been the basis for most of the history and conflict that has transpired on the island over the centuries and that continued into the summer of 1969. In Northern Ireland, over one million people identified their religion as Protestant, while only five hundred thousand were Catholic.[6] The Protestant unionists wanted Northern Ireland to remain part of the United Kingdom, while the Catholic nationalists desired to have the north reunited and return as part of the free Republic of Ireland. The two sides often clashed, with Catholics feeling alienated as they found it more and more difficult to find jobs and housing.

Prior to July 1969, previous marches were marred by 2,300 deaths, some due to rioting, others due to the resistance movement that used terrorist tactics to prevent one side from getting the upper hand. In January, a group known as the People's Democracy, a Catholic loyalist group, began what was called the Long March from Belfast to Derry, modeled after Dr. Martin Luther King's Selma-to-Montgomery March. Only seven miles from their destination, the marchers were attacked by two hundred loyalists throwing bricks and stones and wielding wooden planks, crowbars, and cudgels studded with nails.[7] Dozens were injured and hospitalized, but the violence didn't end there.

In March, two bombs exploded: one did extensive damage to the Holy Family Church in Larne County, Antrim, and the other took out a power substation in Castlereagh, East Belfast, that resulted in the region being without electricity for a number of days. A further series of bombs were detonated across Northern Ireland throughout the summer with the violence increasing on into July at the height of the Marching Season.

On July 12, the rioting in Derry, Belfast, and Dungiven forced families from their homes. Sixty-seven-year old Francis McCloskey died from head injuries after being struck by a baton wielded by an officer of the Royal Ulster Constabulary (RUC). On July 17, Samuel Devenny died after RUC officers had savagely beaten him and his daughters with batons in their home.

All of this eventually culminated in August during what has become known as the Battle of Bogside. The three-day riot began with an annual march to commemorate the Protestant victory in the Siege of Derry in 1689. As the parade passed through the streets, Catholics began hurling stones at marchers. The riot was on.

Rioters hurled Molotov cocktails, injuring many. Police dispensed over one thousand canisters of tear gas. Field hospitals were set up around town. By the third day of fighting, Northern Ireland's prime minister, James Chichester-Clark, took the unprecedented step of contacting British Prime Minister Harold Wilson and requested that the military be brought in.

By August 15, the Bogside riot had ended, but tensions and acts of violence continued. As the tear gas cleared in Derry, over a thousand people were injured as well as 691 police officers. Remarkably, no one died during the rioting.

14

A DEVOTED GROUP OF KENNEDY CAMPAIGN SECRETARIES

They were known as the Boiler Room Girls, six young women who had worked tirelessly as advisors to the presidential campaign of Bobby Kennedy in 1968. Kennedy had been the former US attorney general in the administration of his older brother, President John F. Kennedy, and at the time of his presidential run, was a senator from Massachusetts. Tragically, the younger Kennedy's quest for the presidency was halted by an assassin's bullet while campaigning in Los Angeles.

One of the young women who made up the Boiler Room Girls, Mary Jo Kopechne, had political aspirations of her own. Kopechne was active in the Civil Rights Movement in the late 1950s and early 1960s when she made her first connection with the Kennedy family, volunteering to work on John Kennedy's presidential campaign while she attended college.

Following college, she moved to Alabama, where she taught at a Catholic mission school before moving to Washington, D.C., in 1963 and working as a secretary for Senator George Smithers of Florida. When Bobby Kennedy was elected a New York senator in 1964, Kopechne was hired on to the younger Kennedy's staff as a secretary. During his time in the Senate, Kopechne helped Kennedy write several of his anti-Vietnam war speeches.[1]

In 1968, despite President Lyndon Johnson being the presumed front runner for the Democratic nomination for president, the public had become increasingly disillusioned with the country's involvement in Vietnam. As the months slipped by, Johnson's political standing was eroding. Sensing this, Kennedy made the decision to run for the Democratic nomination against Johnson. Kopechne and five other women joined Kennedy's campaign staff, working endless hours in a small, stuffy, and humid office at Kennedy's campaign head-quarters in Washington, hence the name Boiler Room Girls.[2]

Tragically, Kennedy's campaign lasted only eighty-two days before an assassin's bullet ended his quest in Los Angeles, and the Boiler Room Girls were disbanded.

Fast forward one year to July 19, 1969. It was early in the morning when high school science teacher Robert Samuel and fifteen-year-old Joseph Cappavella lugged their fishing gear out of the trunk of Samuel's car and made their way to the beaches of Massachusetts's Chappaquiddick Island. After hours of surf fishing, the anglers noticed a dark shape in the water. It was the outline of a submerged car.[3]

The pair raced up to a nearby cottage that was being rented out for the summer by Mr. and Mrs. Pierre Malm and knocked on the door. Mrs. Malm answered the knock where the fishermen told her of the submerged car. The two returned to fishing at a different location not knowing what was about to unfold.

Mrs. Malm called the Duke's County sheriff's office at Martha's Vineyard to report the car. By 8:20 a.m., Police Chief Dominik James Arena was on his way to the scene taking the Edgartown ferry the short tenth of a mile across the channel to the island. Upon arriving on the scene, Arena made several attempts to swim to the vehicle, but the current was too strong and he had to wait for help to arrive. Since there had been no reports of an accident, the police chief had a feeling of dread that no one had survived the crash.[4]

As Arena waited for a local diver to arrive, the car's license plate number was radioed back to the sheriff's office for identification. The Oldsmobile Delmont belonged to Senator Edward M. "Teddy" Kennedy.

The diver finally arrived on scene and sank below the water's surface to investigate the crash. The car was upside down on its roof, its engine submerged in the water, the rear wheels protruding skyward. In the backseat he discovered the body of a young woman who was later identified as Mary Jo Kopechne.

Kennedy and several acquaintances were on Chappaquiddick for the annual Edgartown Yacht Club Regatta. Kopechne and the five other women of the Boiler Room Girls were also in attendance and were invited to attend a cookout that one of the Boiler Room Girls, Miss Newburgh, called a gathering for "a devoted group of Kennedy campaign secretaries."[5]

According to the senator's statement to the police, at around 11:15 that evening, Kennedy left the party to return to his hotel. Kopechne joined the senator for a ride back to the hotel she was staying at. En route, Kennedy made a wrong turn down a dirt road and came to a narrow bridge with no guardrails. In the darkness, the car flipped off the bridge and landed upside down in Poucha Pond coming to rest under six feet of water.[6]

Kennedy could not remember how he was able to exit the submerged vehicle and insisted he tried several times to dive under water to extricate Kopechne. Unable to get her out of the car, Kennedy returned to the party and asked his cousin Joe Gargan and friend Paul Markham, to assist. The trio returned to the scene, where they gave up all hope of rescuing the girl after several attempts.

And that's where the story turns and the questions begin. Kennedy continued in his statement that the two men took him to the ferry landing (the ferry had long since ceased operation for the night) and Kennedy swam across the channel where he made his way to his hotel room and collapsed for the remainder of the night. Gargan and Markham returned to the cottage where the party was being held and fell asleep on the floor. None of the three men reported the accident to the police until the morning after the accident at 9 a.m.

Three days after the accident, Teddy Kennedy, wearing a neck brace was joined by his wife, Joan, and Bobby Kennedy's widow, Ethel, in Plymouth, Pennsylvania, to attend Mary Jo's funeral at St. Vincent's Church.[7]

On July 25, Kennedy returned to Edgartown, where he appeared before district court judge James A. Boyle and pleaded guilty to leaving the scene of an accident. The judge gave the senator a two-month suspended sentence and a year probation.[8] The incident was ruled an accident.

That same evening, Kennedy requested television airtime to address his constituents in Massachusetts. During the speech, the senator recounted the incident, defended the young woman's character, and emphatically stated that the two were not having an affair and there was no alcohol involved. He said that his actions following the accident made no sense to him and told viewers that the only explanation was that he was suffering from a concussion and shock.

"No words on my part," Kennedy continued, "can possibly express the terrible pain and suffering I feel over this tragic incident. . . . The grief we feel over the loss of a wonderful friend will remain with us the rest of our lives."[9]

Prior to his speech, public sentiment seemed to have turned against the senator with many calling for his resignation. Kennedy addressed the issue by making a plea to his constituents. "I understand full well why some might think it right for me to resign. . . . I ask you tonight, the people of Massachusetts, to think this through with me in facing this decision. I seek your advice in making it."[10]

Reaction to Kennedy's speech was overwhelming. Not only at the senator's Boston office where the switchboard was flooded with calls that staffers said were 99 percent in favor of his continuing to serve, but at newspapers, radio stations, and television stations as well. An editor at the *Boston Globe* was quoted as saying, "All hell's breaking loose over here. I've never seen it like this."[11] A spokesperson for the Western Union office in Boston said they had fielded 95,000 calls to send telegrams to the senator's office minutes after the speech. Even Kopechne's mother, Mrs. Joseph Kopechne, hoped Kennedy wouldn't resign. "I am satisfied with the senator's statement," she told reporters.[12]

No one knows exactly how many communiques were sent calling on Kennedy's resignation or for him to stay in office, but it is safe to say that those in favor of him staying outweighed the calls for resignation. Boston waitress, Ruth Jarvis, summed the speech up and the feelings of many in the state. "I thought he was terrific," she said. "It could happen to anyone. I hope he doesn't resign."[13]

And he didn't. Senator Kennedy went on to serve in the Senate for another thirty years with one failed attempt at a run for the presidency in 1980. He passed away of a brain tumor in 2009.

While Teddy Kennedy was grappling with the possibility of having to end his political ambitions, another famous politician of the 1960s was reaching the pinnacle of his career. Richard Milhouse Nixon had just been elected the thirty-seventh United States president in 1968 after a turbulent campaign season that saw the assassination of Senator Robert Kennedy and violent riots at the Democratic National Convention in Chicago. Almost immediately following his election, Nixon was under scrutiny by the press, and his mental stability came into question.

Columnist and Washington insider Drew Pearson reported in his weekly column that Nixon was being treated by Dr. Arnold A. Hutschnecker, a specialist in psychosomatic medicine. On July 1, the doctor reported that Nixon was "of superior intellect and keen perception." The doctor went on to say that Nixon would be the most desirable person for the presidency saying that Nixon shows "a controlled reaction when exposed to stress."[14]

From 1969 to 1974, the Nixon administration was littered with scandals, corruption, and crime, which eventually led to his downfall and resignation. But even before Nixon was sworn into office, a murky and questionable plot was uncovered to assassinate the president-elect.

The story begins on November 9, 1968. At the time, Nixon had an apartment at 810 Fifth Avenue at 63rd Street in Brooklyn, New York, but was vacationing in Key Biscayne, Florida. At 10:45 p.m., an anonymous call arrived at Manhattan police headquarters stating that a group of men were planning to assassinate the president-elect. Just over three hours later, New York City detectives and the US Secret Service were raiding the home of forty-six-year-old Ahmad Rageh Namer and his sons, eighteen-year-old Abdo Namer and twenty-year-old Hussein Namer. The older son, Hussein, leapt out of the apartment window and attempted to run down a fire escape but was captured shortly after.[15] The apartment was only five blocks away from Nixon's.

During the raid on the apartment, agents confiscated a .30-caliber carbine, an M-1 rifle, M-1 carbine, twenty-four rounds of .30-caliber ammunition, two switchblade knives, and a "swordlike" knife with a seven-inch blade. The three were booked in a local jail and charged with conspiracy to commit murder, criminal solicitation, and illegal possession of dangerous weapons.

According to a statement by the police, the anonymous caller had demonstrated intimate knowledge of the Namer family and told police that he had been recruited by Namer to carry out the alleged plot against Nixon because "he was an excellent marksman."[16] The call was traced to one of the city's many bars, called Papa's, where the caller was identified by police and later brought in for questioning with the understanding that his identity would be kept confidential.

Rumors began to fly that the family made frequent trips to California and that the trio, because they were Arab, must have indeed been plotting to kill the future president. The reason for the latter rumor was attributed to the fact that only months before, Sirhan Sirhan—an Arab—had assassinated Bobby Kennedy. Surely there must be a connection.

The Secret Service quickly stepped in to knock down reports that the family frequently traveled to California by examining their employment records. In fact, it was found that the three were employed as shipping clerks in a Jewish-owned auto parts manufacturing company. The owner vouched for the men saying that they were good workers and the two cultures had a good rapport with one another. Namer even bought the company owner's wife a bottle of Arabian perfume as a thank you for hiring them.

An announcement as to a possible motive was never released. It was surmised by reporters that since the family was from Yemen, they may have held a grudge against Nixon for his commitment to aiding Israel by arming its troops. A spokesman for the Yemini government jumped in and said the country knew of no reason why Yemenis would be hostile to Nixon.

Five days after the arrests, doubts began to rise about the case against Namer and his sons when Brooklyn District Attorney Elliott Golden told reporters that "there is already some reasonable doubt about the credibility of the informant" who was identified in the statement as Muhammad Hazan Algamal. Later that evening, Golden recanted his statement saying that he was "misunderstood" and didn't want to give the impression there was reasonable doubt.[17] Golden told the press that he "made no such statement . . . nor . . . confirm[ed] any such identity."

After digging into Algamal's background, it was found that he was a disgruntled ex-roommate of the Namers who had thrown him out of the apartment for not keeping it clean. Algamal was later spotted in the Namers' neighborhood where witnesses said he was expressing anger over his treatment.

Yemeni neighbors who were familiar with the Namer family told police that it was common for many Yemenis to purchase weapons in the United States because if they ever had to return to their home country, they would bring the weapons with them to defend themselves in an ongoing civil war in the country.

Eventually on November 16, the three men were released on $25,000 bail, reduced from $100,000. As he left the police station, Namer met with reporters. "I love America. I'm innocent. I have nothing to do against the United States of America. I'm happy in this country."[18]

Finally on July 8, 1969, the Namer trial began. The informant, Algamal, testified before the jury and New York Supreme Court Judge Louis B. Heller that he was in the Namers' apartment prior to his call to the police where the elder of the family urged him to join the plot and that he would be well paid. That is when he went to the bar to make the phone call.

The father, Ahmad, told the jury that he and his sons did not plot to kill the president and Algamal was not in his home when he said he was. Defense attorney David Price argued that Namer and his sons were victims of revenge with the informant being angered by the family throwing him out of their apartment.

The case went to the jury and on July 17, Namer and his sons were found not guilty. They were only found guilty of possessing switchblade knives. And with that, the alleged assassination plot against Richard Nixon was tossed into the dust bin of history.

15

YOUNG MEN WITH UNLIMITED CAPITAL

Hudson Valley towns in New York state such as Woodstock and Wallkill have always been hubs for artists and musicians. As we have seen, as far back as the early 1900s, the beauty and serenity of the valley made it a natural choice for these artisans to come and be inspired. The movement was obliquely an offspring of the Beat Generation, a social and literary movement of the 1950s based in bohemian artist communities of New York's Greenwich Village, Los Angeles, and San Francisco's North Beach area. Members of the Beat Generation felt alienated from the conventional or "square" society and expressed their dissatisfaction with the mainstream through their dress, vocabulary, and political protests. Through heightened sensory awareness provided by drugs, jazz music, sex, or Zen Buddhism, a member of the Beat Generation could obtain personal release from the structured norms of mainstream life.

To be a member of the Beat Generation was to be cool and "hip," but the movement was short lived and quickly faded from the scene in the late 1950s. In the mid-sixties, a new generation began following a similar path and mantra, a feeling of alienation from middle-class society that they believed was based purely on materialism and repression. These young people were labeled with a moniker that was a derivation of the Beat's use of the word "hip." For some it was a badge of honor, while many in mainstream society used it as a term of derision—"hippies."

Professing a subculture of peace and love, these young people were encouraged to "tune in, turn on, and drop out," as one of the celebrity hippies of the day, Dr. Timothy Leary, said in 1966. "Look within [to] find your own divinity. Detach yourself from social and material struggle."[1]

The height of the hippie movement around the world came in 1967 when over one hundred thousand young people converged on the Haight-Ashbury district of San Francisco for what would be the start of the Summer of Love. Throughout those summer months, a heightened feeling of comradery and belonging pervaded the movement with their anthem being the Beatles' song, "All You Need Is Love."

Hippies believed in free love which, to many of their parents, was the ultimate blasphemy. While many times "free love" was meant in the physical sense, for many in this counterculture it was more about the absence of legal ties such as marriage, loving a person for who they are, and not as much about promiscuity.

A crucial part of this subculture was public gatherings. These gatherings could be protests (most notably against the Vietnam War) or they could just be a celebration of life filled with music and poetry. The first such gathering, or Human Be-In, was called the Gathering of Tribes and was held in Haight-Ashbury in January 1967. The event was centered around the Monterey International Pop Festival, a three-day music event that would be the breakout moment for names that would become legendary in the world of rock music—Janis Joplin, The Who, and Jimi Hendrix.

Not all of those who descended on Haight-Ashbury were there for the music. Many were there simply for the utopian experience of peace and love. That utopia was never fully realized, as the moment quickly descended into chaos. The city quickly became saturated and overwhelmed with overcrowding, and with that came unsanitary conditions and crime. As quickly as it had started, by the end of 1967, the San Francisco experiment of the Human Be-In came to an end. A ceremonial funeral was held in October to symbolize the death of the dream, Death to Hippie. The dream may have been dead in San Francisco, but the movement was still flourishing not only in the United States, but across the globe as well.

And that is where four young men—John Roberts, Joel Rosenman, Artie Kornfeld, and Michael Lang—come in. These four men all had the same desire to join in on the youth artistic movement that was burgeoning in the town of Woodstock, New York, which was spurred on by the presence of musical legends such as Bob Dylan and Jimi Hendrix, who often frequented the town.

The quartet were all in their early twenties in July 1969. Roberts was the heir to the Polident denture adhesive fortune. Lang was beginning a successful career as a music promoter. Kornfeld was a songwriter who cowrote the Crispian St. Peters top ten hit, "I'm the Pied Piper," and was a vice president at Capitol Records, the youngest person to ever hold the position. Rosenman was a Yale Law School graduate who was interested in the music business, being a musician himself.

The four were brought together by an advertisement that entrepreneurs Roberts and Rosenman had placed in the *New York Times* in an attempt to find more interesting work after college. They wanted to use their own financing to venture out into unique business opportunities. The ad read, "Young men with unlimited capital looking for interesting and legitimate business enterprises."

Over 1,400 people replied with business proposals including the manufacturing of flying cars and watermelon-flavored popsicles. Along came Lang and Kornfeld. Lang had successfully organized the huge Miami Music Festival the previous year, and of course, Kornfeld had connections in the music world. Together, the men created Woodstock Ventures, Inc., and decided that they wanted to organize a music festival in the beautiful rolling hills of the Hudson Valley and pattern it after the Monterey festival. The proposed date for the festival would be sometime in August.

The men went about signing big-name artists to play the event including Creedence Clearwater Revival, The Who, Jimi Hendrix, Joan Baez, and many more. The only problem was the festival was without a home. The town of Woodstock, which, as we saw earlier, was having issues with "unsavory types," denied issuing Woodstock Ventures permits to hold the concert in the town. Another Hudson Valley city, Saugerties, also denied permits for the festival. Undaunted, Lang, Kornfeld, Rosenman, and Roberts petitioned yet another nearby town, Wallkill, to hold the event there.

Much like all of the towns in the Hudson Valley, Wallkill was a small, close-knit rural community where all of the neighbors knew each other. One resident described the town as being "the kind of place where people still wear McKinley buttons."[2]

Lang and Kornfeld filed an application with the Wallkill Zoning Board to set up the music festival, promising city officials that the event would only bring fifty thousand people to the sleepy town. Ticket sales to that point, however, were proving to be much stronger than anticipated.

When word spread through the town about the proposal, the residents were not pleased. A gas station attendant was asked his opinion by a reporter as he pumped gas into a car. "[It] would be a real catastrophe," he said. "They're going to open the new Sears and Roebuck store then and both those things going on at once would be too much for the town to take."[3]

On July 17, the Wallkill Zoning Board replied to a petition signed by five hundred residents that feared the festival would disrupt town life. In the four-page response, the board stated that the one hundred proposed structures for the event violated local ordinances. The board went on to say that if the event were to be held, the town would hold Woodstock Ventures liable for any zoning violations and would fine the corporation $50 per day for each violation and could face six months in prison for every day there was a violation.[4]

Frank Jennings, the man who organized the petition that circulated around Wallkill, told reporters, "It's not that we're against folk music, jazz, or festivals if planned properly. They didn't plan it right."[5]

Roberts was quick to reply saying that his legal counsel has been "instructed to institute damage proceedings and to provide relief from this offensive harassment and totally dishonest statements of certain individuals."[6]

Hearing of the organizers' dilemma, a local real estate agent directed them to fifty-year-old Max Yasgur's six-hundred-acre farm in Bethel, New York. Yasgur was keen to accommodate the festival for two reasons. First, it was about the money. Though reports vary, Yasgur is believed to have leased the use of his farm for $75,000. Secondly, Yasgur was a believer in bridging the generation gap. "If the generation gap is to be closed," he told *Time* magazine, "we older people have to do more than we have done."[7]

The farmer faced severe backlash from the community. Residents refused to buy milk from his dairy, there were threats on his life, and warnings that his farm would be burned to the ground. Yasgur's wife later said that the threats only stiffened her husband's resolve. The Woodstock Music and Art Fair would go on with over five hundred thousand attending the event over three days in August.

Large gatherings such as the Gathering of the Tribes spread across the United States in July 1969. The July 7 issue of *Time* magazine offered their scathing view of this new counterculture:

> Whatever their meaning and wherever they may be headed, the hippies have emerged on the U.S. scene in about 18 months as a wholly new subculture, a bizarre permutation of the middle-class American ethos from which it evolved. Hippies preach altruism and mysticism, honesty, joy and nonviolence. They find an almost childish fascination in beads, blossoms and bells, blinding strobe lights and ear-shattering music, exotic clothing and erotic slogans. Their professed aim is nothing less than the subversion of Western society by "flower power" and force of example.[8]

Many of the gatherings that sprang up across the country faced the same attitudes that the editors of *Time* professed. Needless to say, the older generation and law enforcement in these cities held similar attitudes and regularly opposed or harassed the attendees of such gatherings. The Crescent City, New Orleans, was one of those cities. Hundreds of hippies first gathered at Mardi

Hand-painted signs like this began popping up across the town of Bethel, New York, when farmer Max Yasgur leased his farm to Woodstock Ventures for a concert promoted as "Three Days of Peace and Love." *Woodstock Whisperer*

Gras Fountain in the center of town for a weekly love-in that brought together like-minded people, free music, and good vibrations.[9] While the love-ins were generally peaceful, except for the occasional swimming and bathing in the fountain, the residents and New Orleans' law enforcement didn't take kindly to the invasion and forced the gatherings to relocate to other parks from week to week. Word of the new location was spread through the city's underground press and by word of mouth.

A reporter for a local underground newspaper, the *In Arcane Logos*, said of the NOLA movement, "Many, many people today are pulling away from the society that produced them. I say to you that we are not going to let the children of the future have rings in their noses and be led around."[10]

Aside from the desire to participate in large gatherings, another aspect of the hippie movement was growing in July 1969—the establishment of communal living. The cover story of the July 18, 1969, edition of *Life* magazine focused on life in one such commune. The headline above a photo of five adults, two women and three men, standing before three young children read, "The Youth Communes: New Way of Living Confronts the U.S."[11]

The concept of communal living where a group of people live together and share responsibilities was not a new concept in the United States. In fact, the idea predates the founding of the nation and throughout the centuries, achieved varied degrees of success. As the 1960s counterculture embraced dropping out of traditional societal norms, they also embraced the concept of communal living, and whether they liked it or not, they brought the concept to the public's attention and made it a part of the 1960s culture.

The *Life* article called these communities "refugees from affluence,"[12] where these young people could flee the crime, squalor, and police harassment of the big city to seek a spiritual rebirth and mutual love, which often times meant more than just loving someone for who they are. It often meant in the physical sense as well with multiple partners. There were communes, including the one profiled in *Life*, where there were married couples living together but not in the legal sense. They believed that they did not need a piece of paper and the blessing of a minister to be legally married, but they still practiced monogamy.

Members of communes, often called "The Family," came from all walks of life—accountants, actors, office workers. Adult members of a commune would share their money, work responsibilities, even the raising of each other's children. Children were educated by the adults and older children. Their religious beliefs were born from any combination of Zen Buddhism, Hinduism, and Christianity.

While it sounds like a utopian way of life free from all of the stresses and burdens of a structured society, life on a commune was difficult, and many members did not last long before returning to the mainstream. Housing usually consisted of makeshift teepees or adobe huts. Those so inclined opted to build geodesic domes. The work was hard, getting up at sunrise to plant, tend, or harvest crops that had to sustain them year-round and doing the nonstop backbreaking work of piling up massive stockpiles of firewood for the brutally cold northeastern winters.

The *Life* magazine article focused on one such commune in Vermont. One of its unnamed members spoke about what the commune experience was all about. "We are entering the time of the Tribal Dance, and as we go to live in teepees, celebrate our joys together and learn to survive. We go to a virgin forest with no need for the previously expensive media of electronic technology. The energy we perceive within ourselves is beyond electric; it is atomic, it is cosmic, it is bliss."[13]

By July 1969, many communes were facing the harsh reality that the self-sufficient lifestyle they sought was not as easy to attain as they thought. The Family profiled in the *Life* article nearly collapsed their first winter when they underestimated how much firewood they would need to survive. Several members had to take part-time jobs so that the commune could purchase food in town. And state law required that their children would have to enroll in a licensed public school in the fall.

It was estimated that over three thousand smaller communes had formed in the United States by 1969. As with all such social experiments, communal life, and indeed the hippie counterculture movement itself, would fade away, but not without a fight. The hippies of 1969 became ever more vocal in the early 1970s in their opposition to the Vietnam War, but by the 1980s, many of those hippies became "yuppies" (Young Urban Professionals) keener on having a professional career and the niceties of life than forging a life away from society.

A FOOTNOTE TO HISTORY

July 14: A fifteen-year-old boy from Lafayette, Indiana, Tom Burns, beat out a field of sixty-five other contestants to take first place in the 1969 Lafayette Soap Box Derby. The win sent him to the national finals in Akron, Ohio, on August 23. The only problem was, as an article in Lafayette's *Journal Courier* said, "Tom is a husky boy" and his win means five weeks of starvation. Tom weighed 120 pounds. His soap box car weighed 130 pounds. That's the maximum total weight for a car and driver to compete in the finals. Tom could not gain an ounce before the competition. His father, Dr. William Burns, said his son would be going on a crash diet before the finals.[14] There is no record of Tom in the official 32nd Annual All-American Soap Box Derby program.[15]

July 18: Famed explorer Thor Heyerdahl and his crew aboard the raft *Ra* were forced to abandon their attempt to prove that it was possible for ancient Egyptians to sail across the Atlantic and influence pre-Columbian cultures in South America.[16] The *Ra* was a duplicate of an ancient Egyptian reed boat. The crew set sail from Morocco on their voyage and were actually scheduled to communicate with the crew of Apollo 11 over short-wave radio relayed to *Columbia*.[17] Instead, the crew found themselves packing what they could and abandoning ship as the *Ra* began to sink six hundred miles short of their destination in Central America.

July 19: One day after Heyerdahl abandoned ship, John Fairfax successfully completed his own trip across the Atlantic, becoming the first known person to ever row across the Atlantic Ocean. The trip began in the Canary Islands and ended in Miami, Florida, a six-month paddle of five thousand miles during which he battled storms, sharks, and exhaustion.[18] The completion of the trip occurred the day before Apollo 11 landed on the moon. Upon returning to Earth, when the crew of Apollo 11 heard of the feat, they sent Fairfax a telegram that read, "We who sail what President Kennedy once called 'the new ocean of space' are pleased to pay our respects to the man who, single handedly, has conquered the still more formidable ocean of water."[19] The explorer would duplicate the feat three years later only this time, rowing across the Pacific Ocean with his girlfriend, Sylvia.

July 19: The Miss Universe Pageant aired on the CBS Television Network live from Miami. From the beginning of the competition, it was clear there was one standout: Miss Philippines, eighteen-year-old Gloria Diaz. Diaz had both beauty and an incredible wit that she wasn't shy to share. The young woman came from a family of twelve children, and when the host asked her to name her two brothers and nine sisters, she replied, "Alphabetically or by rank?" Diaz became the first Filipina winner of the competition and went on to become a successful actress.

JULY 20–26, 1969

Contact Light

16

"KEEP THE VISION . . . KEEP THE DREAM"

At first, it wasn't easy growing up near Oklahoma City, Oklahoma, for a young Jerry Elliott-High Eagle, a Native American whose parents were both Cherokee, but a vision he had when he was only five years old would change his life forever.

Following a difficult divorce, Jerry's mother was forced to fend for her family on her own the best she could. After World War II, she became an executive secretary for the Western Electric Company but barely made enough to make ends meet. The family survived with the aid of food stamps and found that locating a home to live in proved difficult. When they finally found a home, Jerry said it was in a wonderful neighborhood to grow up in and would prove to be an excellent environment to foster big dreams.

"It was a nice neighborhood," Jerry said during an interview with the American Institute for Physics. "Those days, there was no TV and there was radio at night you could listen to. During the day, everyone was outside playing games, doing things, inventing things. We used to make kites out of newspaper and tree limbs and we didn't have to depend upon going to the store and buying anything. If we didn't have it, we made it."[1]

Even though Jerry was Native American, he never felt like he faced any racism in his community or in grade school.

"Grade school was wonderful," he said. "Great teachers, wonderful teachers, great guidance. It was populated with different people—American Indians, white kids. There was no black kids at the time, but at that age, we never looked at racism. We never looked at 'I'm Indian and you're white.' We were all one body of people and I wish we could get back to that."[2]

This was during a time when calculus was still a subject taught in high school and Jerry ate it up along with all of the science courses he could take. He soon realized that a career in physics would be in his future. But his dream actually began much, much earlier on a lazy summer afternoon when he was five and a half years old.

"I was laying outside that house and basically heard a voice saying, 'Your life's work is gonna be landing men on the moon.' I went in real quickly and told mama that and she said, 'Where's the voice coming from?' She opened the door and I pointed to the sun."

Native Americans have deep-rooted beliefs in spirituality and supernatural experiences—visions—that appear to offer guidance and protection. Jerry's mother was a firm believer. "She nodded and she said, 'Keep the vision. Keep the dream.' From the age of five and a half, I knew I was gonna land men on the moon."

The problem was that there was no NASA at the time. In fact, there was no space program when the vision came to Jerry, and who knew when, if ever, there would be such a program.

While in middle school, the only thing Jerry cared about was science and math. He was so bored with social studies that he would often just sit in class and endlessly lob wads of paper from the back of the classroom into the waste basket next to his teacher's desk. Many times he would be caught and sent to the principal's office.

"Thinking back . . . I could have explained it," Jerry chuckles. "'Well, gee, in my future, I'm gonna be a Retrofire Officer at the NASA Mission Control Center that computes return-to-Earth trajectories and all I'm doing is practicing."

The years passed, and Jerry never lost sight of the vision. He went on to become the first Native American to graduate with a degree in math and physics from the University of Oklahoma. Not long after, he applied for a position with NASA, but in the interim, he was drafted into the military to go fight in Vietnam. With only fifteen days left before he was scheduled to leave for boot camp, a telegram arrived at Jerry's home that read, "Dear Mr. Elliott. We are offering you a position in the Man in Space program. Please call immediately."

Jerry called NASA's personnel director Bernie Goodwin and informed Goodwin that he would not be able to accept the offer. He had been drafted.

"Who's your draft board director?" Goodwin asked.

"It's Colonel Williams, Sir," Elliott replied.

"Don't worry," Goodwin said. "We have General Stevenson on our staff that would be happy to call Colonel Williams and tell him you are ours."

And that was it. Jerry Elliott-High Eagle's career with NASA began. In short order, he was in the Mission Control Center in Houston manning a console for the flight of Apollo 11, the first Native American to do so. He was also on the team that helped guide Apollo 13 back from the moon after the explosion that nearly cost the country the mission and the lives of three astronauts.[3] And it all began with a vision.

17

T PLUS FOUR DAYS

As Apollo 11's command module *Columbia* and lunar module *Eagle* entered lunar orbit on July 19, they found that they had company—the Soviet Union's Luna 15 spacecraft had arrived at the moon two days prior.

For two days after its arrival, engineers back on Earth studied possible landing sites for the unmanned Luna 15 spacecraft, which was scheduled to land on the moon and return rock and soil samples before Apollo 11. What Luna 15's engineers and scientists found was a harsh and rugged landscape that would make for a tricky and dangerous landing. The decision was made to have the spacecraft do two course corrections that would, with luck, put the lander over a better landing location.

The first course correction took place on July 19. The second occurred right on time on July 20 just before 9 a.m. eastern time. The Russians made the decision that when they found a suitable landing site, they would cut the amount of time the spacecraft was on the moon to only two hours, just enough time to pick up a few samples then liftoff for the trip back to Earth. But even after the two course corrections, the Russian scientists found that the new landing sites were just as foreboding as the first. Flight controllers once again delayed Luna 15's landing another eighteen hours. In the meantime, Armstrong and Aldrin had begun their own descent to the moon.

The lunar module *Eagle*, with Neil Armstrong at the controls, also had a challenge with the landing. The lunar module was designed so that the landing would be partially automatic, partially manually controlled by the astronauts, but as was the case with Luna 15, the area that was selected as a landing site in the Sea of Tranquility was boulder strewn, so an automatic landing was out of the question. Armstrong immediately took the matter into his own hands and took control of the spacecraft.

One of the reasons Armstrong was selected for this mission was because of his coolness under pressure. He had demonstrated this trait to NASA twice before. The first occasion was during the flight of the two-man Gemini 8 mission in March 1966. The plan for that mission was for Armstrong and his crewmate David Scott to rocket into orbit and catch up and dock with an unmanned Agena target vehicle that would be launched just prior to the astronauts from Cape Kennedy. The Agena would be used by this and future Gemini missions to prove and practice docking procedures that would later be required to connect the lunar module and command module during Apollo missions.

Not long after the two astronauts had docked successfully with the Agena, they suddenly found themselves spinning out of control end over end. Armstrong made the decision to undock from the Agena, but after the two craft separated, the spin continued, increasing rapidly to the point that the two astronauts were on the verge of blacking out.

As the spin reached one revolution per second, Armstrong realized that the spin was being caused by a thruster on the Gemini capsule that wouldn't shut off, causing the spacecraft to roll end over end. Armstrong was able to turn the thrusters off and use the spacecraft's reentry control thrusters to stop the spin and return the crew safely to Earth.[1]

The second time was during a practice session for the Apollo 11 landing in the Lunar Landing Research Vehicle at Ellington Air Force Base near Houston. The vehicle was nicknamed the "Flying Bedstead" due to its spindly legs. The multi-engine craft was the only method astronauts had that would allow them to practice landing the lunar module, which was not designed to be flown here on Earth.

Armstrong had flown the vehicle twenty-one times before without incident, but on the twenty-second flight, things went wrong. The vehicle began rocking out of control. Armstrong did his best to correct the motion, but with only seconds remaining before crashing into the runway, the astronaut ejected only two hundred feet above the ground. The vehicle exploded on impact while Armstrong floated safely to the ground by parachute.[2]

As *Eagle* was descending toward the lunar surface, an alarm bell broke the concentration of all involved—it was alarm 1202, but what was a 1202 alarm? No one in mission control had ever heard of it. Was it an alarm that would jeopardize the landing?

There was silence from mission control as they scrambled to find out what this unknown alarm was. Finally, a twenty-seven-year-old software engineer,

Jack Garmin, remembered that he had seen that alarm only once before during a dress rehearsal for the landing. It was *Eagle*'s onboard computer telling the astronauts that it was being overwhelmed with data. Mission control radioed up to the astronauts that they were still "go" to continue with the landing.

Minutes later, Armstrong faced another problem. They had overshot their intended landing site by four miles and were heading into a boulder field. As he did with Gemini 8, the astronaut switched off the automated guidance and began to land the spacecraft manually. Both Armstrong and Aldrin peered out of *Eagle*'s windows looking for smooth terrain to land. At this point, the lander only had sixty seconds of fuel remaining and a crash landing could be in their future.

Aldrin radioed back to Earth, "Kicking up some dust."

Mission control chimed in, "Thirty seconds . . ." (the amount of fuel remaining).

Aldrin then made the announcement, "Contact light."[3]

After a series of tasks were completed to save the lander, making sure that its engine was properly shut down, Armstrong radioed back to Earth, "Houston, Tranquility Base here. The *Eagle* has landed."

With fuel tanks almost dry and a possible collision with boulders avoided, the relief expressed by capsule communicator, astronaut Charlie Duke, said it all: "Roger, Tranquility. We copy you on the ground. You got a bunch of guys about to turn blue. We're breathing again. Thanks a lot."[4]

The astronauts ran through a post-landing checklist, ensuring the lander's engines and other systems were properly shut down or oriented for a quick escape from the moon just in case, then moments later, Buzz Aldrin performed another first in human history.

Long before this moment when Aldrin learned that he would be part of the first lunar landing, the astronaut approached the head of the astronaut office, former Mercury astronaut Deke Slayton, with a request. As soon as he and Armstrong landed on the moon, Aldrin would like to hold a communion service to give thanks for the incredible achievement. Slayton bristled at the idea, not because he was against it, but because NASA and the agency's administrator, Thomas Paine, were embroiled in a lawsuit that was brought against them following the flight of Apollo 8.

On Christmas Eve 1968, the crew of Apollo 8—Frank Borman, Jim Lovell, and Bill Anders—became the first humans to fly to the moon. Although they didn't land, it was still a monumental achievement in human history. Their iconic photo titled *Earthrise* that showed the blue marble of our home planet in the inky blackness of space rising before an equally drab and lifeless moon was so moving and inspiring, that it helped give rise to a massive worldwide environmental movement.

Knowing that millions of people would be watching their adventure a quarter of a million miles away from Earth on this sacred holiday, the crew made the decision to send a Christmas greeting back to the people of Earth. They read from the Book of Genesis.

That is when devout atheist Madalyn Murray O'Hair filed the lawsuit against NASA and Paine. In her brief, she contended that the reading from the Book of Genesis violated the Constitution's First Amendment and the separation of church and state.[5] Hesitantly, Slayton granted approval for Aldrin's service but advised the astronaut that millions of people across the globe with wide-ranging religious beliefs would be watching. He advised Aldrin that he should keep his comments less controversial.

The astronaut agreed and the moment Armstrong uttered those now immortal words, "the *Eagle* has landed," Aldrin began to hold communion on the moon. Before silencing communications for the service, he made a brief statement to the world. "I'd like to take this opportunity to ask every person listening in, whoever and wherever they may be, to pause for a moment and contemplate the events of the past few hours and to give thanks in his or her own way."[6]

Communications were cut for a few minutes as Aldrin took out a golden chalice that had been loaned to him by the minister of his church, Reverend Dean Woodruff, as well as a communion wafer, and a small vile of wine and began the service.

"I poured the wine into the chalice our church had given me," Aldrin later recalled. "In the one-sixth gravity of the moon, the wine curled slowly and gracefully up the side of the cup. It was interesting to think that the very first liquid ever poured on the moon, and the first food eaten there, were communion elements."[7]

Upon their return to Earth and after decontamination, the chalice was returned to his church, the Webster Presbyterian Church in Houston, where every year since the historic landing in 1969, the church brings it out to hold a special ceremony marking the anniversary of the moon's first communion.[8]

Two hours after landing, *Eagle*'s hatch was opened and Neil Armstrong scaled down its ladder, lighting upon *Eagle*'s footpad where he described what the lunar soil looked like. At 10:56 p.m. on July 20, Armstrong stepped off the footpad.

"That's one small step for [a] man . . . one giant leap for mankind."

For a brief moment in human history, the world was unified. In the United States, over fifty-three million people watched Armstrong take that first step on

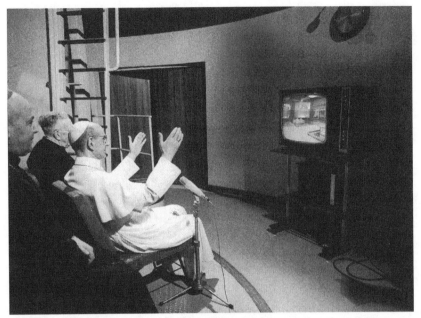

Pope Paul VI watching Neil Armstrong's first step on the moon from the Vatican.
European Space Agency

television in their living rooms, restaurants, churches. Worldwide, that number was estimated to be over six hundred million. Tears flowed and cheers filled the air in cities and towns around the world.

In Spokane, Washington, American flags dotted homes that lined city streets. An Air Force sergeant stationed in town saw the displays and commented, "This is the type of demonstration I like to see," a reference to the multitude of protests that had occurred over the past decade.

In a Miami bar, a patron contemplated what the moment meant. "The moon is the only place in the universe where there is both humans and peace."[9]

Leafy Noble of Moonville, Indiana, reminisced about a time fifty years prior. "It was even more exciting than that day in July 1918 when the first airplane flew over Moonville."[10]

Millions of people around the world celebrated the feat, many feeling the pride as if they were Americans that night. In Nigeria, Sarah Maxson and her husband at the time were Peace Corps volunteers, two of only a handful of American teachers in the country. The couple along with other volunteers from different countries took chairs outside and sat in the brilliant starry night, a shortwave radio tuned to the Voice of America.

"[We] marveled that we were hearing voices speaking from the bright crescent overhead," Maxson recalled. "Many friends joined us as we toasted the
astronauts with Star Beer. To our surprise, over and over, people shook our
hands and congratulated us on this amazing feat, as though we were responsible
for it."[11]

Soon after Armstrong took mankind's first step on another world, Buzz
Aldrin hopped down the ladder to join his partner. Taking a look at the landscape that unfolded around them, Aldrin framed the scene perfectly, calling it,
"magnificent desolation."

Once on the surface, Armstrong and Aldrin began their mission by first
deploying a television camera so that mission control and people on Earth
could watch their every step. Next, they began to deploy several experiments,
unfurled the American flag, and revealed the commemorative plaque attached
to the lunar module's landing leg. They also accepted the longest long-distance
phone call ever, a call from President Nixon in the White House's Oval Office.

In all, the astronauts would spend twenty-one and a half hours on the moon.
Two hours before their scheduled departure, the Russian Luna 15 began its
final descent to the lunar surface, but only seconds later, communications with
the lander was lost and Luna 15 was never heard from again. Engineers believe
that it had crashed into a mountain at 298 miles per hour. The Soviet's attempt
to land a man on the moon was defeated when their N1 moon rocket exploded
on the launch pad only days earlier. Now Luna 15 had been destroyed. The
Russians' quest for the moon had ended.[12]

Besides the American flag, the plaque that was attached to the lunar module,
and several other commemorative items that would be left behind, Armstrong
and Aldrin had a special package to deliver. It was carried by Aldrin in a pocket
just under one of his shoulders. In that package were mementos that paid hom-
age to the astronauts and cosmonauts who made the ultimate sacrifice for the
exploration of space. The first was the Apollo 1 mission patch with the names
of the three astronauts—Gus Grissom, Ed White, and Roger Chaffee—embroidered on it. The three men perished in the first manned test of the Apollo
capsule during a dress rehearsal on the launch pad in 1967. It also contained an
identical set of gold olive branches that were presented to the Apollo 1 wives.

The package also contained two Soviet medals. The first honored Vladimir
Komarov who died in 1967 aboard the Soyuz 1 spacecraft. The other was to
honor the first man in space, Yuri Gagarin, who tragically died in an airplane
accident in 1968. Along with these tributes, the astronauts were to leave behind
the silicon disc that contained the messages of goodwill from world leaders
etched on it.

Keep in mind that the two astronauts were on the moon for a truly short time—twenty-one hours, thirty-six minutes total—and they were incredibly busy, so it's understandable that they might forget to complete a task or two. One of those they almost forgot was leaving that package on the moon. It nearly made a return trip home to Earth.

At 111 hours, 36 minutes into the flight, Armstrong and Aldrin were about to reenter the lunar module and prepare for liftoff from the lunar surface to rejoin Collins in *Columbia* for the ride home. Aldrin was halfway up the LM's ladder and was about to enter the spacecraft when Armstrong remembered something.

"How about that package out on your sleeve?" he asked Aldrin. "Get that?" Aldrin replied, "No."

"Okay," Armstrong replied. "I'll get it when I get up there."

In a later interview, Aldrin pointed out that the astronauts knew what they had forgotten and without being specific, he knew exactly what Armstrong was talking about. The pair didn't want to discuss it openly. On television screens back home, Armstrong makes a silent move to his right. Aldrin had tossed the package to the ground and Armstrong had pushed it slightly with his foot to make sure it was secure on the moon.

"Okay?" Armstrong asks Aldrin about the placement of the package.

"Okay," Aldrin replies.[13] Mission accomplished. Package delivered.

Climbing in and out of the lunar module was quite the challenge when wearing a bulky spacesuit equipped with a huge backpack life-support system. As Armstrong later commented, "It's very cumbersome. You're like the Frankenstein monster."[14]

As Aldrin was entering the LM, his backpack hit a special switch, the *ENG ARM* switch. This little button would activate *Eagle*'s ascent stage engine so that the astronauts could ignite it and lift off the moon to rendezvous with Collins, who was waiting in the command module. If the switch tripped prior to liftoff, there would be no engine ignition, and the astronauts would be stranded on the moon.

Years later, Armstrong joked with reporters, "He could have picked something that was not very important."[15]

Both Armstrong and Aldrin were confident that the switch would not trip, but they wanted a little reassurance. Since they had jettisoned all unnecessary gear from the spacecraft onto the lunar surface to lighten the load for liftoff, they used the only "tool" they had left to secure the switch for a little added insurance—they wedged a Fischer Space Pen into the button that activated the switch thus ensuring that it wouldn't trip like a faulty circuit breaker in a home power box and the crew could leave the moon.

At 1:54 p.m. on July 21, *Eagle*'s ascent-stage engine fired and performed flawlessly, hurling Armstrong and Aldrin from the moon's surface where they would rejoin Collins in *Columbia*. Four hours later, the lunar module was jettisoned and the large engine attached to *Columbia*'s service module ignited, setting the capsule and crew on a return trajectory back home to Earth.

18

WHAT A SPECTACLE!

The journey home for the crew of Apollo 11 was relatively uneventful but unbeknownst to many, the actual splashdown in the Pacific Ocean could have quickly turned into a tragedy due to a tremendous thunderstorm nicknamed a "screaming eagle" that was heading toward the landing site. The looming catastrophe was picked up by two military meteorologists stationed at Pearl Harbor, Hawaii. The problem was that they were not affiliated with NASA and were watching the storm unfold through the eyes of a top-secret military satellite that the world, even NASA, knew nothing about. They were not authorized to divulge the information.

At the time, NASA was reliant on a division of the US Weather Bureau (now called the National Weather Service) and the Spaceflight Meteorology Group (SMG) for weather information. Both agencies' forecasts were highly reliable, and as splashdown was drawing near, they noticed some disturbed weather heading toward the Apollo 11 recovery area that was producing twenty-knot winds, six-foot seas, scattered showers, and isolated thunderstorms, but nothing that would prevent splashdown. The weather was well within splashdown parameters.[1]

Meanwhile, Air Force meteorologist Hank Brandli, who was stationed at Hickam Air Force Base at Pearl Harbor, was monitoring one of the government's top-secret spy satellites that tracked the movements of the nation's enemies—Cuba, China, and the Soviet Union. In the images, Brandli saw the towering, billowing clouds of the brewing storm expanding out from a barely visible center. He knew immediately that a tropical storm was forming, and it was heading toward the Apollo 11 recovery area.

The satellite that Brandli was monitoring was much more advanced than the ones in use at the time by NASA and the Weather Bureau, but it was top secret, and the images were classified, so he couldn't inform NASA directly of the impending doom—a storm strong enough to shear Apollo 11's parachutes causing the capsule to plummet into the ocean, killing the three-man crew.

"I knew Apollo 11 would come back and they would get killed because I had this classified information," Brandli told *Weatherwise* magazine.

Brandli contacted a go-between who was also stationed at Pearl Harbor, Navy meteorologist Willard "Sam" Houston. Houston was tasked with relaying weather forecasts to the fleet of ships that would recover the Apollo 11 astronauts and had the necessary clearances to see the images. He agreed with Brandli's assessment, but now the hard part. He had to convince the commander of the recovery fleet, Rear Admiral Donald C. Davis, a difficult task because Davis did not have authorization to view such material.

"[He] had to convince Admiral Davis without the photos, which were from a satellite that wasn't supposed to exist," Brandli said. "He couldn't tell him how he knew what he knew."[2]

The information was relayed to the NASA flight director and, not wanting to take a chance, the call was made for the carrier USS *Hornet* and supporting ships to reposition themselves. Mission control radioed to the Apollo 11 crew that they would have to perform a maneuver that would take the capsule away from the storm.

"I wasn't happy with that fact," command module pilot Michael Collins was later quoted as saying. "The great majority of our practice and simulator work and everything else had been done [to a different] target point."[3]

It was the right call, and on July 24, eight days after their launch from Cape Kennedy, the Apollo 11 capsule with its crew and cargo of nearly fifty pounds of moon rocks and soil was whizzing their way through space at 9,671 feet per second, heading for a landing just past the massive storm. The crew prepared for the final phase of the flight—their fiery return through the Earth's atmosphere and a splashdown nine hundred miles southwest of Hawaii in the Pacific Ocean.

The only way the general public could view the splashdown was on their television sets with images relayed from the recovery vessels. The closest anyone could get to see the event in person was if they were on board the recovery aircraft carrier USS *Hornet* or were a passenger on a very special Qantas Airline flight.

Qantas is the official overseas airline of Australia. The airline became famous worldwide in the 1960s with their television commercials that featured a disgruntled koala bear complaining about the number of tourists the airline was bringing to the country invading his home. The koala ended each commercial with the tagline, "I hate Qantas."

Qantas pilot Captain Frank Brown was scheduled to pilot flight 596 from Brisbane, Australia, to San Francisco, California, the same day Apollo 11 would return. Brown and Qantas management realized that the flight would have a most astounding view of the spacecraft's reentry, a view that no one else on Earth would have.

Weeks before the flight, Brown began checking his flight path daily to make sure that there were no deviations that would prevent the passengers and crew from seeing the reentry. Their flight would fly approximately 280 miles parallel to the incoming capsule.

On splashdown day, flight 596's takeoff was delayed for two hours to ensure that the aircraft could make its rendezvous with Apollo 11. As the flight cruised high above the Pacific Ocean, twelve-year-old passenger Trevor Hiscock said his eyes were "as big as the portal windows" as a glowing streak appeared in the dark sky.

In the cockpit, Captain Brown came over the plane's loudspeaker to address the passengers. His words were also linked to the coverage of the splashdown being aired by the Australian Broadcasting Company.[4]

Good morning, ladies and gentlemen. This is Captain Brown speaking from the flightdeck of Qantas flight 596. It does seem we are going to get a very good look at Apollo 11. I think I . . . Here they come! On the left side! Two of them! One object brighter than the other. See the two of them, one on top of the other, the brightest one lowest? One's the command module, one's the service module. Each weighs six tons. They're just picking up heat now. The bottom one is leaving an incandescent trail. [Do] you see them flashing? That's a trail of epoxy additive resin coating. What a spectacle![5]

In honor of this special flight, and maybe to placate some of the passengers for the delay in its departure, Qantas offered passengers a commemorative Apollo 11 certificate while first-class passengers could dine from a special menu that included Duckling à la Armstrong, Roast Loin of Lamb Aldrin with Splashdown Sauce, and Chilled Lunar Lobster with Collins Salad. To top off the meal, everyone received a Moon Mint.[6]

At 5:44 a.m. local time, the spacecraft's drogue parachutes deployed to slow the capsule down, followed a short time later by three gigantic orange-and-white main parachutes that gently floated the crew to splashdown in the Pacific Ocean only six minutes later, ending the flight of Apollo 11, but the mission itself was far from over.

When NASA first began ironing out the details about how they would fulfill President Kennedy's lunar ambitions in the early 1960s, a debate began among scientists about whether or not the moon's extremely thin atmosphere could support any lunar life, even down to the smallest microbe. They also pondered what effects introducing Earth germs to another world would have on the moon. Most scientists believed that the moon was void of life, but some, including a young Carl Sagan, were not so sure and asked one simple question: How do we know that there is no life there at all?

To err on the safe side, engineers devised a system to quarantine the astronauts, their gear, and the lunar samples to prevent them from contaminating the Earth with moon germs. The process began immediately after splashdown when the hatch would open and the crew would don special biological suits complete with canister-type respirators before being flown to the aircraft carrier and sequestered inside a sealed containment unit, basically a small Silverstream-looking trailer, that had all of the comforts of home in a small, cramped area. The astronauts would remain in the trailer until it was delivered to the Manned Spaceflight Center in Houston, Texas, where they would transfer into a specially designed Lunar Receiving Lab (LRL). The LRL was designed with several biologically sealed rooms. No germs could get in or out of the building. The crew would quarantine in the LRL for twenty-one days.

Sounds like a—no pun intended—airtight plan, but as astronauts Michael Collins and Buzz Aldrin later pointed out, it had a few drawbacks. In an interview with the PBS series *American Experience*, the two mentioned that moon dust sticks to everything, so when the explorers returned to the orbiting command module after working on the moon, despite their best efforts to remove the dust from their spacesuits, it would still escape into the capsule. After splashdown and once the Navy recovery divers opened the hatch, that dust could conceivably be exposed to the Earth's atmosphere.

After donning their biological suits, Navy divers in a raft next to the bobbing capsule would scrub the astronauts down with rags dripping with disinfectant. When finished, the divers would tie a weight to the rags and drop them into the ocean where they would sink to the bottom. Any moon germs that didn't die from the cleaning would now be in the ocean.

Fortunately for the world, scientists were correct in their initial hypothesis that the moon was a lifeless rock. Actually, NASA found out fairly quickly that the moon dust returned on the flight did not contain any life forms and those findings came quite by accident.

The crew of Apollo 11 were flown from the bobbing raft in the Pacific Ocean by helicopter to the waiting recovery ship, where they walked a few yards from

Inside the Mobile Quarantine Facility, Apollo 11 astronauts (left to right) Michael Collins, Edwin E. Aldrin Jr., and Neil A. Armstrong relax following their successful lunar landing mission. *NASA*

the aircraft to the Mobile Quarantine Unit wearing their biological suits and looking much like aliens themselves that the suits were meant to protect the Earth from.

Inside the relatively comfortable but cramped quarters, the three opened a curtain in the rear window of the unit and were able to peer out to the throngs of sailors and press that had gathered around. The Apollo 11 trio gathered at the small window, cramming together so that they could be greeted by President Nixon, who was on board the carrier to personally welcome them home.

After a short speech and several questions from the president including looking for a reply to an invitation that he extended to the crew to join him for a state dinner in Los Angeles (although Nixon alluded to the fact that the astronauts' wives had already committed to them being there), the ship's chaplain, Lt. Commander John A. Piltro led those in attendance in prayer, praising the crew and the staggering accomplishment that had just been achieved, but also reminding the world that humans can achieve anything they set their mind to.

"As we tried to understand and analyze the scope of this achievement for human life," Piltro began, "our reason is overwhelmed with bounding gratitude and joy, even as we realize the increasing challenges of the future. This magnificent event illustrates anew what man can accomplish when purpose is firm and intent corporate.

"A man on the moon was promised in this decade, and though some were unconvinced, the reality is with us this morning in the persons of the astronauts: Armstrong, Aldrin, and Collins."[7]

From Hickam Air Force Base in Hawaii, the quarantine trailer containing the astronauts and an entourage of medical staff was loaded aboard a C-141 Starliner aircraft and airlifted to Ellington Air Force Base in Houston, where a crowd of 2,500 people, including the astronauts' wives, had gathered at two in the morning to greet the returning heroes. Although still inside the trailer, Janet Armstrong, Joan Aldrin, and Pat Collins and their children were able to talk to the men via a telephone connected to it.

The trailer was then transported to the Lunar Receiving Laboratory where the astronauts, their lunar samples, and a contingent of scientists and workers began moving the men and material in for their three-week stay.

Inside the LRL, sequestered behind the building's biological barrier along with the astronauts, was NASA photographer Terry Slezak. One of the photographer's jobs was to remove film magazines from protective canisters that traveled to the moon with Apollo 11. Those magazines contained the hundreds of photos that Armstrong and Aldrin had taken while on the moon.

As he started pulling the magazines from the containers, Slezak found a handwritten note from Aldrin that read, "This is the canister that Neil dropped on the surface but this is the most important magazine."[8]

Not fully realizing the meaning of the note, Slezak pulled the magazine from the canister—bare handed—and saw that it was covered in a thick black material. Then he noticed that the material was now on his hands. It was lunar dust. The photographer had inadvertently become the first man to physically touch the moon. Slezak had to immediately strip down, clean his work area with bleach, shower, then start his own quarantine period.

Besides the possibility of lunar microbes, a more severe problem with lunar dust was that it is like volcanic ash. It's made up of sharp, glasslike particles that could cause severe even deadly respiratory issues if inhaled. But being the consummate professional, Slezak was more concerned with the film than his own health. As mentioned, those particles get into every nook and cranny and stick to anything and everything. It could have scratched the film and made it

Terry Slezak in the Lunar Receiving Lab showing Moon dust on his hand from the Apollo 11 film canisters. *NASA*

unusable. After three weeks, Slezak showed no signs of having a reaction to touching the dirt. Both the photographer and film were OK.

With moonrocks in hand, scientists began examining the material and performing experiments on them, which included subjecting a number of biological subjects to the Apollo lunar material. They ground up moon rocks and mixed it into aquarium water that contained live fish to see if they had a reaction. They fed portions of the ground-up material to cockroaches and insects. They injected moon dust into plants, Japanese quail, shrimp, even oysters. The scientists were basically trying to verify the effects that any possible microbes from the lunar samples might have on a representative sampling of living species on Earth and avoid a worldwide plague. Scientists even performed one of the age-old scientific experiments—inoculating a chicken egg with a sample.

A short time after Apollo 11's return to Earth, a story appeared on the back pages of the *New York Times*. The headline read, "Woman Pricks Finger in Test of Moon Dust."[9]

Twenty-two-year-old Nancy Klein was running experiments with lunar dust. The dust was contained in a special hermetically sealed cabinet that had rubber gloves built into its side where researchers could reach in to run experiments and examine the material without fear of contamination.

Mrs. Klein was inoculating chicken eggs with a syringe filled with moon dust. The needle slipped and penetrated not only her glove but also her finger making her the first person to be vaccinated with moon dust.

Following protocol, Mrs. Klein was placed into quarantine to make sure she did not become ill from the incident. After three weeks, Nancy Klein was given a clean bill of health and released from quarantine. Not long after, the process of quarantining astronauts ended following the flight of Apollo 14.

Aldrin, Collins, and Armstrong would remain in quarantine for three weeks, during which time they would be debriefed by NASA officials and scientists from behind protective glass walls, but the most grueling part of their mission was yet to come.

19

NOT EVERYONE HAS GONE TO THE MOON

The list of celebrities, politicians, and world leaders who wanted to attend the launch of Apollo 11 was enormous. As the chief of the astronaut office Deke Slayton said of the event, "Everybody and his brother wanted to be at the launch."[1]

Recognizing this, NASA invited over twenty thousand VIP and IIPs (Incredibly Important People) to the Kennedy Space Center to have the best seats for the show on July 16. NASA public affairs officer Gene Marianetti was assigned the task of lining up drinking water, security, and portable toilets for the guests at the Cape's viewing areas. His job also included herding twelve thousand of those guests onto buses on launch day to a special viewing area only three and a half miles away from the rocket, while another six thousand were issued special parking passes so that they could have VIP parking near one of the next best viewing positions located a bit farther away but still on the grounds of the space center.[2]

The night before the launch, a small VIP dinner party was held at the Cocoa Beach Country Club[3] with the likes of famed aviator Charles Lindbergh, US Army General William Westmoreland (who had just returned from Vietnam), and the man whom CBS News anchor Walter Cronkite called "The Father of the Program"[4] for championing NASA and space exploration well before President Kennedy made the commitment to land a man on the moon before 1970, former President Lyndon Johnson.

The following day saw television talk show host Johnny Carson and his sidekick Ed McMahon, comedian Jack Benny, former president Johnson, and current vice president Spiro Agnew in the bleachers. Additionally, 69 ambassadors from foreign countries, 100 foreign science ministers, 19 governors, and 275 business leaders were in attendance.[5] And those numbers did not include the estimated one million people who lined the beaches, parking lots, and causeways surrounding the space center or the countless number watching the event at home live on TV.

Taken at face value, it appeared that the world had Apollo 11 fever and that everyone had gone to the moon, and, yes, for a fleeting eight days in July 1969, the entire world seemed to be unified with the flight of Apollo 11. Well before the historic launch on July 16, 1969, however, there was a strong undercurrent, a push back on the government for spending what was viewed as an incredible sum of money on a flight of fancy instead of spending it on social and environmental issues back on Earth.

Gallup polls throughout the 1960s had shone a spotlight on this fact. With the exception of one year, 1965, polls consistently showed that while Americans had a favorable view of NASA and the astronauts, they were less than enthusiastic with spending billions of dollars on the moon landings. Favorability ratings for all NASA manned projects consistently hovered between 30 and 50 percent with the majority of those who responded indicating that they would have rather seen the money spent on social projects like the environment and eliminating poverty.[6]

The two main factors for this attitude stemmed from the Vietnam War, which was becoming more and more unpopular as it dragged on endlessly, and the millions of Americans who were living in poverty. While the poverty level in 1969 decreased slightly from the previous year, down from 25.4 million people to 24.3 million, it was still a staggering number.[7]

With that in mind, Congress began chipping away at NASA's budget from $5 billion in 1965 to $4 billion in 1969,[8] forcing the agency to shelve its grand plans for manned exploration following the Apollo missions in favor of a single, reusable space shuttle that would operate in Earth orbit.

In 1966, President Johnson's own director of the Bureau of the Budget, Charles Schultze, sent a memo to the president requesting that NASA funding be cut and that the moon landings be pushed back to the 1970s so that all of the president's "Great Society" programs could be funded.[9] Johnson refused to budge on the issue.

The following year, Schultze sent another memo urging Johnson to make cuts in the program. This time, Johnson reluctantly agreed but under the condition that those cuts would only take effect after the Apollo program was completed.[10]

Environmentalists were also talking about the costs involved in flying men to the moon. One year prior to the Apollo 11's launch, the *New York Times* ran an editorial titled "To Save Spaceship Earth." Spaceship Earth was the moniker the planet was given after a photo taken by the crew of Apollo 8 one year earlier titled *Earthrise* showed just how fragile the Earth was in the darkness of space. In the editorial, the *Times* made an accurate comparison between the Apollo capsule and Spaceship Earth:

> When the astronauts orbit the heavens, they are acutely aware that their contin-ued existence depends upon the limited supplies of oxygen, water, food, and other essentials contained in their capsule.
>
> This planet—which some call "Spaceship Earth"—also has limited resources, and as these approach exhaustion or are made unusable by pollution, the margin of safety for all men declines.
>
> The parochial priorities implied by this attitude must deepen concern as to whether Spaceship Earth can be saved for its proliferating cargo of fragile organ-isms whose increasingly ravenous demands steadily deplete and damage the ecological basis of life on this planet.[11]

Critics of the Apollo program throughout the 1960s ran the gamut from famed anthropologist Margaret Meade, who hoped for the "disciplined and courageous use [of the space program's] enhanced human powers for man,"[12] to philosophers like Lewis Mumford, who wrote, "Space exploration . . . is strictly a military by-product; and without pressure from the Pentagon and Kremlin, it would never have found a place in any national budget."[13]

Washington Post columnist Drew Pearson wrote about a nationwide water shortage the country was experiencing, comparing the shortage with the Apollo 11 mission saying that the "two illustrates man's efficiency in achieving the thrills of life and man's inefficiency in not achieving the necessities of life." Pearson continued:

> At Cape Kennedy, the United States is about to launch the most carefully rehearsed, most expensive, most unnecessary project of this century by which man will reach a piece of drab, radioactive, lava-like real estate hitherto roman-tic because of distance—the moon. The launching will succeed because a vast amount of money and the best scientific brains in America over a period of seven years have been lavished on this moon shot. Meanwhile, up the Atlantic Coast, the Capitol . . . is desperately short of the second essential to man's life—water— all because of a lack of planning, lack of foresight, and lack of money—the same ingredients which have put the moon shot on the verge of success.[14]

As the CBS television network prepared to cover the first moon landing, network executives made the decision to bring into their coverage opposing views, people who believed that the moon landing was not worth the cost. The head of the network's news division, Gordon Manning, decided the two people they needed to put on the air as a counterpoint to the historic landing were feminist organizer and activist Gloria Steinem and renowned author Kurt Vonnegut.

As CBS News producer Sandford Socolow recalls, Manning rushed into his office the night of the landing:

> He said, "We've got to get someone on here who's going to say this is nonsense. Call Gloria Steinem." Two o'clock in the morning and we send a limousine to pick her up and bring her in. Somebody else at Gordon's behest had gotten Kurt Vonnegut to come in. And so there we had a panel of Kurt Vonnegut and Gloria Steinem telling us how there were so many more important problems on Earth and what were we doing going to the moon?[15]

The reporter most affiliated with America's space program, Walter Cronkite, was anchoring the coverage. "I remember those two particularly who were bitter about the fact we were on the moon," Cronkite remembered in an interview with the Television Academy. "[They said] we should have been taking care of the poor on Earth. We had 'em on the air *that very day* [the day of the landing]. And incidentally, they got a tremendous amount of insulting mail. I would suggest because it seemed that they were raining on our parade."[16]

The feeling that money should be diverted from the space program to other pressing social issues was a worldwide sentiment. As Apollo 11 began its journey, over in Britain, science fiction author Ray Bradbury was on a panel with several other guests during the BBC's coverage. Also on the panel was Irish political activist Bernadette Devlin. Devlin was raised by her mother in poverty after her father passed away when she was nine. Her youth in the ghettos of Ireland helped form her socialistic views, one of those was about spending too much money on space and not on Earth.

During the telecasts, Devlin made her views quite clear to the audience, so much so that Bradbury finally turned, looked the woman straight in the eye, and said, "This [Apollo 11] is the result of six billion years of evolution. Tonight, we have given the lie to gravity. We have reached for the stars. . . . And you refuse [to] celebrate? To hell with you!"[17]

The one group of Americans who felt the most disenfranchised by the space program were African Americans. In 1969, the poverty rate for Blacks was at 31.1 percent compared to 9.5 percent among whites.[18] Three years prior to the launch of Apollo 11 and two years before his death, Reverend Dr. Martin Luther King Jr. addressed the issue during testimony before the senate subcommittee investigating the plight of inner cities. His opening statement was direct and to the point:

> Beyond the advantage of diverting huge resources for constructive social goals, ending the war would give impetus to significant disarmament agreements. With the resources accruing from termination of the war, arms race, and excessive space races, the elimination of all poverty could become an immediate national reality. At present, the war on poverty is not even a battle, it is scarcely a skirmish.[19]

The headline in the July 21, 1969, edition of the *Chicago Defender* blared in big bold type, "World Stands Still: Moon Shot Unites U.S. for Instant." The article went on to say that it was the first "non-racist event in American history."

"At this moment, people of every race, nationality, age and condition were united in praise for an achievement symbolic of the American genius. This Sunday was the unexpressed sentiments of millions of Black Americans."[20] In reality, that was not the consensus.

On July 19 as Armstrong and Aldrin were preparing to set foot on the moon, the *New York Times* sent reporters out across the city to get a sense of the community's reaction to the event. In Harlem, fifty thousand people turned out for the Harlem Cultural Festival at Morris Park. When Apollo 11 was mentioned by a speaker, a tremendous chorus of boos rang out. In a bar, a patron lamented, "There ain't no brothers in the program where they can get some of that big money. . . . I don't like them saying 'all good Americans are happy about it.' I damn sure ain't happy about it."[21]

In an effort to get all Americans on board with the landing of *Eagle* on the moon, President Nixon made the executive decision to proclaim July 21, 1969, a federal holiday calling it a "day of participation" so that all federal employees could watch the landing and moon walk and encouraged the business community and schools to do the same. Former Student Non-Violent Coordinating Committee (SNCC) chairman and future Washington, D.C., mayor Marion Barry responded by calling for a boycott of the holiday adding, "Why should blacks rejoice when two white Americans land on the moon when white America's money and technology have not yet even reached the inner city?"[22]

This view had been festering beneath the surface in Black communities across the country for years. Three years before Apollo 11 was scheduled, seventy-five protestors staged a sit-in beneath a mock-up of the lunar module at the Manned Spaceflight Center in Houston demanding the government provide affordable and clean housing for those in need before another dime was spent on space missions.[23]

The US space program finally met its detractors in the Black community face to face on July 15, the day before the launch of Apollo 11. It was a meeting that occurred in a field just outside of the gates of the Kennedy Space Center.

Reverend Ralph Abernathy had taken up the reins of the Southern Christian Leadership Conference (SCLC) and the Poor People's Campaign following the tragic assassination of his close friend and mentor, Dr. Martin Luther King Jr. Abernathy had come to the forefront of the movement in 1963 when he and Dr. King organized and led the Poor People's March on Washington in an effort to shine a light on the employment inequalities and housing issues facing the poor across the country.

On July 12, Abernathy and the field director of the Poor People's Campaign, the Reverend Hosea Williams, announced that "hungry people from five Southern states" would demonstrate at Cape Kennedy July 15 and would try to get as close as possible to the launch site with mules and wagons.

"We're not against things like the moon shot," Williams told reporters. "But there has been a miscalculation in priorities."[24] It was announced that Reverend Abernathy would lead the procession.

The afternoon before the launch, a light misty rain was falling as a group of five hundred protestors marched to a field at the Kennedy Space Center, the procession being led by a mule-drawn wagon, a stark contrast with the gleaming white Saturn V rocket that was bathed in the brilliant light of giant searchlights. The protestors held signs reading, "Moonshots Breed Malnutrition," "Rockets or Rickets?," and "Billions for Space, Pennies for Hunger."[25] Abernathy himself stood at the gates holding his own sign that read, "$12 a Day to Feed an Astronaut. We Can Feed a Starving Child for $8."

Concerned that the protestors might block the one roadway that led to the space center thus preventing access by key NASA personnel, former president Johnson questioned NASA Administrator Thomas Paine about security. Paine assured Johnson that he would speak with the protestors and address their concerns personally.

With a small group of reporters in attendance and the protestors singing "We Shall Overcome," Paine arrived at the gate where the protestors were waiting at 3 p.m. Following standard government protocol, Paine wrote a Memorandum for Record to put on file that described the meeting:

> The Reverend Abernathy gave a short speech in which he deplored the conditions of the nation's poor, stating that although he had no quarrel with the space program,

it represented an inhuman priority at a time when so much suffering exists in the nation. One fifth of the population lacks adequate food, clothing, shelter, and medical care, he said. The money for the space program, he stated, should be spent to feed the hungry, clothe the naked, tend the sick, and house the shelterless.[26]

Paine looked at Abernathy and said, "If we could solve the problems of poverty in the United States by not pushing the button to launch men to the moon tomorrow, then we would not push that button. The great technological advances of NASA were child's play compared to the tremendously difficult human problems [on Earth]."[27]

Paine then asked Abernathy to "hitch his wagons to our rocket, using the space program as a spur to the nation to tackle problems boldly in other areas" and that he would "do everything in my personal power to help in his fight for better conditions for all Americans."

Abernathy accepted Paine's response and prayed for the three men aboard Apollo 11 before addressing reporters. "On the eve of man's noblest venture, I am profoundly moved by the nation's achievements in space and the heroism of the three men embarking for the moon. What we can do for space and exploration, we demand that we do for starving people."[28]

The Reverend then made three requests of Paine: that NASA support the movement to combat the nation's poverty, hunger, and other social problems; that NASA scientists use their talents to tackle the problem of hunger; and that ten families from the group of protestors be allowed inside the space center to view the launch.

The next morning, a bus arrived at the gate and loaded a group of the marchers onboard so that they could view the launch from the VIP seating area inside the space center. On board, Abernathy's marchers found breakfast and a candy bar waiting for each of them on the seats. In his Memorandum for Record, Paine wrote, "I wished it were as easy to meet his other two requests."[29]

This sentiment among Americans that the moon landing was a frivolous expense continued through the remaining years of the Apollo program. One notable voice was that of Black poet and songwriter Gil Scott-Heron who told his followers that the lunar missions were distractions from the country's Earthly issues. "[It's] just something to hold down the pressure and revolt in America."[30] In response to Apollo 11, he wrote a poem that focused on the frustrations and reality of life in urban America–"Whitey on the Moon":

A rat done bit my sister Nell,
with Whitey on the moon.

Her face and arms began to swell,
and Whitey's on the moon.
I can't pay no doctor bill,
but Whitey's on the moon.
Ten years from now I'll be payin' still,
while Whitey's on the Moon.

Eventually, NASA did modify some of its Apollo technology for use in inner cities across the country including converting space-age insulation like that used in the capsules into energy-efficient windows for public housing, new technology that would measure air pollution in urban areas, and innovative technology to create fresh drinking water.

20

"IT'S 'PLAY POWER'"

It had been building since 1963 when the Black residents of York, Pennsylvania, began peaceful protests over discriminatory practices in City Hall. Throughout the year, demands to make the police force biracial had been ignored by the all-white city council. In response, the residents took to the streets, where they were met with violence from police attempting to rein in the protestors.

It was during this time that all-white street gangs began to appear in York, including one called the Girarders. For the next six years, the gangs terrorized the Black community with street brawls and random fire bombings of homes and businesses.

By July 1969, the violence had crescendoed to the point of no return. On July 17, a young Black man accidentally burned himself while playing with lighter fluid but blamed the accident on the Girarders. That same day, seventeen-year-old Taka Nii Sweeney was shot by an unknown gunman after being stopped by a York police officer for violating the city's youth curfew. The match had been lit, and the streets of York filled with Black youths and the white gangs. The two groups began throwing rocks at one another and taking pot-shots with guns. The melee lasted into the following day. By the end of the second day, eleven people had been injured, one fatally—twenty-two-year-old Officer Henry Schaad. Schaad had only been in the York police department for eleven months when he was shot. He died two weeks after the shooting. The violence didn't stop. Entire city blocks were set on fire, and by July 20, one hundred people had been arrested and over sixty people injured.

On July 21, a young Black woman from Aiken, South Carolina, Lillie Belle Allen, was in town visiting her parents. Allen's sister, Hattie Dickson, was driving Allen through town when she took a wrong turn as they searched for a grocery store to buy some supplies. As the car turned the corner, Dickson spotted guns pointing out of the second story windows of the buildings lining the streets. It was the Girarder gang.

Dickson stepped on the gas pedal and attempted to turn the car around, but it stalled out in the middle of the street. The armed white men poured out of the buildings. Allen jumped out of the car and begged the men not to shoot but to no avail. It was reported that over one hundred shots were fired at the car. Allen was struck multiple times and died.

Immediately, Governor Raymond Shafer declared a state of emergency. Two hundred National Guardsmen rolled into the city and an emergency curfew was put in place. Three days later, the rioting finally ended.

For the next two years, detectives tried to locate and bring to trial the men who were involved in the killing. "It was tougher than pulling teeth," lead detective Thomas Chatman Jr. told reporters. The townspeople didn't want to rat out one of their own or were living in fear for their own life.

The case went cold for the next twenty-eight years before a new interest in the events of that sweltering summer night in July 1969 began to come together. Suddenly, clues as to who murdered Lillie Belle Allen began springing up. Several of the gang members involved had committed suicide. Donald Altland, who was a member of one of the gangs at the time of the shooting, was a suspect and was questioned by police but did not admit to the killing. Altland later committed suicide, but before dying, he left a taped confession and a message scribbled on a napkin that read, "Forgive me, God."

By the spring of 2001, ten men had been arrested in connection with the murder including York's mayor, Charles Robertson, who was a police officer at the time of the shooting. According to testimony, Robertson had given a gang member ammunition and told him to "kill as many niggers as you can."

Seven of the defendants pleaded guilty to lesser charges. Two were found guilty of second-degree murder, while Robertson was acquitted.

The case to find the people who murdered Officer Schaad was also reopened in 2001, and Stephen Freeland and Leon Wright were brought to trial and found guilty of second-degree murder.

In a study conducted by President Johnson in 1967, the National Advisory Commission on Civil Disorders released a report that reviewed the causes for the violence occurring in cities across the United States such as York. The report was blunt and straight to the point:

Segregation and poverty have created in the racial ghetto a destructive environment totally unknown to most white Americans. What white Americans have never fully understood but what the Negro can never forget is that white society is deeply implicated in the ghetto. White institutions created it, white institutions maintain it, and white society condones it.[1]

Many of the issues that were igniting cities across the United States in July 1969 stemmed not only from racial tensions, but also a stream of frustrations: the continuing fight for equitable and habitable housing for lower-income families, feeding the poor, and the fight for basic civil rights and liberties. By 1969, protests had become the norm, both peaceful and otherwise. They offered people the chance to raise their voices in solidarity for a cause.

On July 12, a call went out across Indianapolis, Indiana. Reverend Andrew J. Brown, chairman of the Indianapolis chapter of the Southern Christian Leadership Conference (SCLC) asked that "organized labor, businesses, churches, youths, government officials, ghetto residents, and Negro moderates and militants"[2] join him for a march to the governor's mansion and a rally with the intent of focusing attention on the plight of the poor in the community and, in particular, force the governor to reconsider measures that were either dropped or vetoed by the state house that dealt with civil rights, welfare, and aid to dependent children. The march would wind along the streets of the city through a Black neighborhood where over five thousand homes were taken away for the construction of Interstate 65, as well as a mixed-race neighborhood and an all-white neighborhood, before gathering at the mansion.

"People who scream about violence should get together and work with the ones who want to settle the problems by non-violence," Marion County AFL-CIO president Max Byrdenthal told reporters. "It must be settled by one or the other: violence or non-violence."[3]

The following day, six hundred marchers and speakers including the Reverend Jesse Jackson assembled and began the three-and-a-half-mile walk to the mansion. Governor Edgar Whitcomb was not at home when the marchers arrived. After several speeches and without provocation, approximately half of the group began climbing the walls surrounding the governor's house. Reverend Brown addressed those breaching the mansion's gates by reminding them that this was supposed to be a nonviolent march and promised them that the Reverend Ralph Abernathy, SCLC's national chairman, would come to the city to help resolve the issues at hand if needed as long as they did not damage any property.[4]

State police arrived on the scene and with that, the marchers disbanded quietly but not before breaking windows to the tune of $300 in damages. Due to the episode, Governor Whitcomb made it clear in a July 29 press briefing that he would not meet with any of the demonstrators, stating that it would not be logical to discuss business or government under the force of demonstration, but would be "willing to meet with the people of Indiana to discuss state matters at the state capitol any time."[5]

Whitcomb believed that there was something more nefarious about the march to the mansion. In his press briefing, Whitcomb laid out his belief that six busloads of marchers were not actually from Indianapolis but Chicago and that their goal was to establish a "program of mob violence" instead of actually addressing the plight of the poor. The governor suggested that the money spent sending the busloads of marchers to Indiana would have been better spent in Chicago where "there is much more poverty."[6]

The governor also addressed the participation of Republican state representative David Allison, saying that his appearance at the rally was a grab for publicity. He then dismissed any further talk of the incident, introduced the newly crowned Miss Indiana, Jill Jackson, and changed the subject.

The civil rights fight for African Americans in the 1960s had been a long and often perilous road. There were victories and violence, but those victories, no matter how small, were historic, including the July 28 special election in Greene County, Alabama. Five seats on the county commission went up for grabs when the US Supreme Court ordered the election after they heard arguments that six African American candidates had been intentionally left off the general election ballot the previous November.

In one of the largest turnouts in county history, four thousand voters went to the ballot box, over twenty-five hundred were Black. In the end, four of the five seats were won by Black candidates. Probate Judge J. Dennis Herndon called it a "black day for Eutaw [the county seat]."[7]

SCLC chairman Reverend Ralph Abernathy came to Eutaw to celebrate the event with a prayer service. "I couldn't miss this!" he said.[8]

While much of the focus of historians has been on the Black civil rights marches of the 1960s, another group was coming to the fore demanding equal rights and the end of violence against their community—the gay or what would later be called the LGBTQ+ community. One of the most notable events to occur in the movement's history came in June and July 1969 at the Stonewall Inn.

The Stonewall Inn was a bar located almost dead center in New York City's Greenwich Village district on Christopher Street only a few doors down from the offices of the legendary newspaper, the *Village Voice*. Patrons would either grab a drink with friends in the front room or take to the back room where quarter after quarter was pumped into a jukebox so everyone could dance the night away. The Stonewall became a popular gathering spot for gays and lesbians. It was a safe haven where they could socialize without fear of retribution or condemnation. As the founder of the Gay Liberation Front and Stonewall patron Jerry Hoose recalled, "The bar itself was a toilet, but it was a refuge. It was a temporary refuge from the street."[9]

While homosexuality was legal in New York State in the 1960s, in a sense, the State Liquor Authority declared that bars that served alcohol to gay patrons were "disorderly houses." Seeing a new market, in 1966, a member of the Genovese crime family, Tony "Fat Tony" Lauria, purchased what was until then a straight bar and turned it into a gay bar. Lauria made a killing by controlling the money that flowed from the sale of cigarettes, watered-down drinks, and the jukebox while at the same time being able to keep the establishment open by bribing New York's Sixth Precinct police department with $1,200 a month to turn a blind eye.

The Stonewall was a dive in the truest sense of the word. There was no running water, and many times patrons drank from dirty, previously used glasses. The building only had one exit in the event of a fire—the front door.[10] Despite all of this, the bar flourished. Patrons were just looking for one place where they could be themselves. The Stonewall was it.

Even though the police were being paid bribe money, they would still regularly raid the Stonewall. Police officers would burst into the bar, turn on the lights, then begin to line up patrons and check their identification. If a person did not have an ID, if men were dressed in drag, or women did not have on three articles of feminine clothing, they would be arrested.[11]

Since the establishment opened in 1966 as a gay bar, patrons just succumbed to these raids without a fight. That all changed early in the morning of June 28, 1969, as eight police officers entered the bar. Inside, two hundred people chatted and danced the night away. On this night, the patrons stood their ground and refused to cooperate with the officers. The officers made the decision to arrest everyone in the building, but since the paddy wagons hadn't arrived yet, everyone was made to wait inside.

A few were told they were not under arrest and stepped outside onto the street where they waited to see what would happen next. By the time police backup arrived, passersby and angry residents had joined the group waiting outside the club, which now numbered in the hundreds. A lesbian woman was handcuffed and dragged to a waiting police wagon. When she complained that the cuffs were too tight, she was hit on the head with a baton and thrown inside the waiting van. The crowd had finally had enough.

"We've had all we can take from the Gestapo," one patron commented to a *New York Daily News* reporter. "We're putting our foot down once and for all."[12]

In the ensuing rage, the crowd began throwing bottles and bricks at the officers. One of the police wagons was overturned. Parking meters were uprooted and used to smash the Inn's windows. The building itself was set on fire and a tactical unit had to be called in to extricate the officers who initiated the raid and the patrons who remained inside.

The riot was finally brought under control at 4 a.m., three hours after it had started. But that wasn't the end of it. News spread like wildfire through the Village and the following night, the riot picked up where it left off. Shops were looted and fires were lit.[13] For the next four nights, over one thousand protestors showed up and the violence continued until eventually it ended on July 1.

Thirty days after the riot had ended, on July 30, the gay rights movement took to the streets of New York once again. Five hundred marchers gathered at the city's Washington Square fountain, all wearing lavender shirts, smocks, and armbands in a show of solidarity. They planned to march through the city streets to the Stonewall Inn.

Before the march began, a voice would shout over a megaphone, "Give me a 'G'!" The crowd would respond with a rousing "G!" The voice continued: "Give me an 'A'!" The crowd replied. "Give me a 'Y'!" Again, the crowd responded with the letter. The spelling lesson continued: "P-O-W-E-R! What's that spell?"

"Gay Power!" the crowd roared.

A New York City tour bus had pulled up next to the throng of marchers and listened to the chants. A man leaned out of one of the bus windows wielding a home movie camera to capture the sights of the city. A woman next to the man asked, "What are they shouting, Henry?"

"It's 'Play Power,'" he replied. "It's a demonstration."[14]

At the Washington Square fountain, a spokesperson for rally organizers, Martha Shelley, picked up a megaphone. "Brothers and sisters," she began. "Welcome to the city's first gay-power vigil!

"We're tired of straight people who are hung up on sex," she continued. "Tired of flashlights and peeping-tom vigilantes. Tired of marriage laws that punish you for lifting your head off the pillow. Socrates was a homosexual. Michelangelo was a homosexual."[15]

The marchers headed out and made their way to the Stonewall where they broke into song singing "We Shall Overcome." As a *Village Voice* reporter put it, "Gay power had surfaced. Sick and tired of police harassment . . . and vicious busts. Homosexuals had struck back. [The July 30 march was] a mild protest to be sure, but apparently only the beginning."[16]

The Stonewall Riot was a galvanizing moment for the LGBTQ+ community. It was the first time gays and lesbians stood up for themselves and their rights. One year after the riot, thousands of people took to the New York City streets yet again, this time to peacefully march in the Christopher Street Liberation Day Parade, the first gay pride parade in the country. The parade continues to this day on the anniversary of the Stonewall Riot, and offshoots are held in cities across the country.

In 2016, President Barack Obama designated the Stonewall Inn and surrounding district of the Village a national monument. It was a small way of honoring those who had made a huge contribution to the gay rights movement.

21

A MIRACLE IN
THE MAKING

Baseball has always been "America's Pastime." Since its earliest days, the game has been played in some form or another under many different names (Round Ball, Stick Ball, etc.) with variations on the rules. It is generally recognized that the game took on more or less the form we know today back in 1839 when Abner Doubleday laid out the first plan for a baseball diamond in Cooperstown, New York,[1] although some have questioned whether or not Doubleday had actually codified the game.

In 1939, the baseball world held a yearlong centennial celebration of Doubleday's "invention" during which baseball greats of the day, including Joe DiMaggio, Ted Williams, and Hank Greenberg, wore specially commissioned patches on their uniforms that featured a blue diamond, a ball player wielding a bat embroidered above a white baseball, and the words "1839–1939 Baseball Centennial" emblazoned atop a red-and-white background. The celebration was also highlighted with the opening of the National Baseball Museum in Cooperstown.

So 1939 is recognized as the year when the concept of the game was set (for the most part) in stone. But when did it become a professional sport? Sports historians trace the year to 1869, when the Cincinnati Red Stockings became the first team to announce publicly that they were paying their players' salaries. Not only were they the first team to pay its players, but they also did the unthinkable—they went undefeated with a record of sixty-seven wins and no losses. So it was that 1869 was recognized as the birth of professional baseball, and it was only fitting that 1969 should have its own centennial celebration.

The celebration took on many forms, the first being the creation of a new logo for the newly founded Major League Baseball Promotion Corporation. Up until 1968, there was no corporation overlooking the sport. In that year, the moniker, "Major League Baseball" (MLB) was first used. The corporation's new logo was released in 1969, the iconic red-and-blue logo with the white silhouette of a player at bat in the center, which is still in use to this day.

The league commissioned an official record album to commemorate the event. *Baseball: The First 100 Years: Official Centennial Record Album* was produced by Fleetwood Records and incorporated audio highlights from the sport's past and narration by actor Jimmy Stewart and sportscaster Curt Gowdy.

The league also sponsored a poll asking fans to rank the greatest baseball lineup of all time. The results would be announced by Apollo 8 astronaut Frank Borman at what has been described as "the grandest ball ever thrown by baseball." The ball was held at the Washington Sheraton Hotel on July 21, the night before the annual All-Star Game, with twenty-three hundred people attending including thirty-four all-star players, the largest gathering of baseball greats ever at one time.

Over the course of human history, the word *miracle* has been tossed around quite freely, and the summer of 1969 saw an uptick in its usage. The fact that NASA was nearing the finish line to reach President Kennedy's ambitious goal was a miracle in and of itself, considering all of the technical and human challenges that the agency faced along the way and in such short time—only seven years since the president made that bold declaration. There were some events that were not as monumental as a moon landing but were still touted as being a "miracle," one of which was taking shape on a baseball diamond in the summer of 1969 in the contest over the National League East championship.

As the heat of summer continued to rise, it looked as though the Baltimore Orioles would be unstoppable in the American League Eastern Division, as they held a comfortable eleven-game lead over the Boston Red Sox. In the west, the Minnesota Twins and Oakland A's were deadlocked in first place. Over in the National League West, the Los Angeles Dodgers clung to a slim half-game lead over the Atlanta Braves.

Meanwhile in the National League East, the Chicago Cubs led the New York Mets by eight games. Now to many readers not familiar with the Cubs and Mets you may think "so what?" But for the Cubs to even be in first place was a miracle in and of itself. The team had not won the World Series in sixty-one years, the last time being in 1908. In the interim years they suffered one disappointing, heartbreaking loss after another and Cubbie fans knew why—it was the Curse of the Goat.

The curse began in 1945, the last time (until 2016) that the Cubs were in the World Series. In that World Series, Chicago held a lead of two games to one over the Detroit Tigers. A local resident, the owner of the Billy Goat Tavern, Billy Sianis, bought two tickets to game four, one for himself and one for the tavern's namesake goat. You read that right. He bought a ticket for his goat so that they could both watch the game together. The ushers at Wrigley Field refused to let the goat in. Sianis was incensed and was allegedly heard to say, "Them Cubs, they ain't gonna win no more!"

And they didn't. Chicago went on to lose the series in seven games and until the 1969 season, they were odds-on favorites to miss the Fall Classic once again, but this season was different. As July 1969 rolled around, manager Leo Durocher led the Cubs to a record of fifty wins and twenty-eight losses and were sitting comfortably eight games ahead of another team that fans had written off for the season, the New York Mets. But if you weren't a believer in curses and superstitions, what happened next surely changed your mind, and it was more than just that goat that jinxed the Cubs.

The New York Metropolitans, or "Mets," first took to the field in 1962 at the brand new Shea Stadium in Flushing, New York. Flushing is a bustling neighborhood of Queens where Jewish, Irish, German, and Italian cultures blend to create a vibrant community. It was the site of not one but two World's Fairs, one in 1934, the other in 1964. For the latter, a giant steel Unisphere depicting the Earth was the centerpiece and was ringed with forty-eight fountains. The new stadium in Flushing was the site for the first major stadium concert ever held when the Beatles performed there to a hysterical crowd of over fifty-five thousand in 1965.

The Mets entered the National League in an attempt to fill the void left behind by the Brooklyn Dodgers and New York Giants when they moved to Los Angeles and San Francisco, respectively, in 1957. For baseball fans, New York had been the epicenter of the sport, with the triad of the New York Yankees, Brooklyn Dodgers, and New York Giants once making up a powerhouse of teams. They ruled the game in the 1940s and 1950s, with the Yankees usually coming out as winners of the World Series. The Mets, however, were a sad replacement for the two teams that had left for the West Coast.

Since their inaugural season in 1962 to this point in the 1969 season, the Mets had finished in last place every year. Their debut season saw them finish with a 40–120 record, placing them 60 games out of first place. Their record didn't improve much over the ensuing years, but this year, 1969, seemed different, primarily because they were winning games fairly consistently. So with the Mets actually winning games and the Cubs sitting comfortably in first place, it was understandable that fans of these two teams were rightfully anticipating that a miracle was about to occur.

On July 8, the fortunes of one of the two teams would begin to change. The Cubs were in Flushing for the start of a three-game series against the Mets. It was the bottom of the ninth inning of the first game which was being played in the afternoon. The Cubs were ahead by a score of 3–1 and were three outs away from winning the game and padding their lead over the Mets.

The Mets brought second baseman Ken Boswell in as a pinch hitter for their pitcher, Jerry Koosman. Boswell hit the ball into centerfield, but the sun at the stadium at this time of day made it difficult to see a ball in the sky, and the Cubs' Don Young missed making the catch. Boswell was in with a double.

After Mets' centerfielder Tommy Agee fouled out, pinch hitter Donn Clendenon blooped another pitch into centerfield, and again Young missed making the catch. Clendenon was on base with a double, while Boswell had to hold up on third base.

Another of the Mets, Cleon Jones, promptly doubled to right field allowing the two base runners to score. The game was tied at 3–3. After intentionally walking the next batter and forcing Wayne Garrett to ground out, first baseman Ed Kranepool came to bat. With a one-ball, two-strike count, Kranepool singled to centerfield. Jones scored from third base and the Mets won.

The jubilation seen on the field was something that is routine in the game today, but in 1969, it was a first, an emotional outpouring that was unheard of. The Mets cleared the dugout and swarmed Kranepool, piling on him in a mass of celebration. It was a turning point for both teams.

From that moment on, the Mets continued their drive toward first place as the Cubs began to falter. Later that summer on September 9, in a crucial game between the two teams at Shea Stadium, a fan dropped a black cat onto the field. The cat ran out onto the diamond, looked at the Mets' dugout, then at the Cubs' dugout. Both teams watched anxiously to see what the cat would do. At this stage of the season, it was normal to believe in superstitions and this one could spell doom for one team or the other.

The cat finally made up its mind and ran in front of the Cubs' dugout. From that game to the last pitch of the regular season, Chicago went on to lose their last eighteen out of twenty-seven games, and the "Miracle Mets" were headed to the World Series, where they defeated the Baltimore Orioles for the crown and their first championship.

Off the ballfield on July 22, more musical icons were making news. In the Detroit, Michigan, suburb of Highland Park, the "Queen of Soul" was having a moment when she was arrested for disorderly conduct. Franklin was involved in a minor parking lot accident which was responded to by two metropolitan police officers.[2] No one is sure what was said to offend the singer, but she began swearing at officers before trying to slap one of them. She was arrested and sent to the local police station where she posted a $50 bail and was released. As she drove off from the police station, she ran down a traffic sign.

One of the groups that had made a name for themselves at the wildly successful Monterey International Pop Festival in 1967 and who had just signed on to play at Woodstock the following month, The Who, was having an incredibly prolific and successful summer.

Earlier in the year, the British band was holed up in IBC Studios in London beginning to record one of the seminal albums of the 1960s, *Tommy*. In March, the first single from the album was released, an ode to the fanatical love of a table-top game, pinball, called "Pinball Wizard." Today, the song is one of the most recognizable and beloved by the band, reaching number 4 on the British Billboard charts but only number 19 in the United States.

The full album was released on May 17, 1969, and was billed as being the world's first rock opera. The double-album set tells the story of a deaf and blind mute who suffers unthinkable abuse as a child but eventually overcomes his past to become a pinball champion. The boy's name was Tommy, a common British name during World War II, much like "Joe" was for Americans during the war.

"What it's really all about," lead guitarist Pete Townshend said of the album, "is the fact that [Tommy] is seeing things basically as vibrations, which we translate as music. That's really what we want to do: create this feeling that when you listen to the music you can actually become aware of the boy, and aware of what he is all about."[3]

In June, Townshend met with executives from Universal Studios in Los Angeles who offered the band a two-picture deal, one of which would be a full-length adaptation of Tommy. On July 19, Townshend made the announcement that the movie would be made with a $2 million budget and that he would be working with a scriptwriter to bring the album to the big screen.

It would take six years for the movie to be completed, but when it was released on March 19, 1975, it becomes a huge commercial success. The movie starred a who's who of entertainment giants—Tina Turner, Eric Clapton, Elton John. The Who's lead singer, Roger Daltrey, played the lead role with actress Ann-Margret portraying his mother. Daltrey, Ann-Margret, and Townshend were nominated for Academy Awards for their performances and the music.

The album became the number 1 selling album in 1969 and has gone on to sell twenty million copies over the years, making it the eighty-fourth-biggest-selling album of all time.

In the world of television, children's programming was about to take a giant leap from mindless cartoon characters to something that was fun but educational at the same time. A show was about to debut that would teach children through the use of fast-paced cartoons and an incredible lineup of characters without the young viewers even knowing they were learning something.

The show was the brainchild of a Public Broadcasting Service (PBS) documentary film maker, Joan Ganz Cooney, who had the vision of creating programming for preschoolers and underprivileged three- to five-year-old children to get them ready for school. In 1966, she began drawing up the idea for what would be called *Sesame Street*.

In July 1969, test production began on the show that would take place on a set resembling a New York City neighborhood with a back alley, brownstone houses, trash cans, small shops, and a community of friendly neighbors. It would combine funny live segments with short cartoons that would keep a child's (and adult's) attention while at the same time teach them something—math, emotions, spelling, the subjects ran the gamut.[4]

The first test before a live audience of children did not go well. Most were not happy with the not-so-cheery set. Enter Jim Henson, creator of the Muppets, who came up with a character to brighten up the scenery—a giant yellow bird with orange legs that would have a childlike quality about him. His name would be Big Bird. That one change brought new life to the set, and from that, a complete cast of Muppet characters were developed to join the live actors—the trash-can-living Oscar the Grouch, best friends Bert and Ernie, who shared a basement apartment, and another Muppet character who was already making a name for himself, Kermit the Frog.[5]

The theme song was written by Joe Raposo and was an upbeat, fun song with lyrics that one of the show's creators, Jon Stone, thought was "a lyrical disaster" with "hackneyed phrases." Stone thought lines like "everything's A-OK" and "sweeping the clouds away" would quickly become obsolete. And here we are in 2023. The theme remains and sticks in the head of everyone who hears it.

Production continued into August, and the final result, the first episode of *Sesame Street* premiered on November 10. The series has never looked back, only developing over the years to fit its ever-changing audience.

Washington Post cartoonist and reformed smoker Herbert L. Block created this image depicting the US House of Representatives as being a "filter," a blockade between the Federal Trade Commission and the Tobacco Industry which was being required to remove advertising from television and radio. *Courtesy Herbert L. Block Collection, Library of Congress (004.17.00) LC-DIG-hlb-07291*

A FOOTNOTE TO HISTORY

July 22: The annual Major League Baseball All-Star Game was to take place this evening at RFK Stadium in Washington, D.C., and air on NBC television at 7 p.m. The game, however, was rained out and rescheduled for the following afternoon. It was the last time the game was ever rained out and the last time it was played in the afternoon. The National League beat the American League 9 to 3.

July 23: Earlier in the month, a six-year congressional prohibition against placing mandatory health warnings on cigarette labels and the abolition of cigarette advertising on radio and television was allowed to expire. Following the lifting of this prohibition, cigarette manufacturers faced extreme pressure from both Congress and the Federal Trade Commission to "do the right thing": warn the public about the health hazards of smoking and no longer glorify the habit in on-air advertising. On July 23 at a hearing in Washington, the cigarette manufacturers yielded to the pressure and agreed only under the condition that the new and permanent prohibition would not begin until 1970.[6] Congress and the FTC agreed, and that was the beginning of the end for the Marlboro Man in television ads.

July 24: "Refreshing Mints" are introduced to the world. Contained in a small clear plastic case, the mints made a now famous clicking sound when the package was shaken. The product was soon renamed, "Tic Tacs."

July 25: Sharon Sites Adams made her first visit to Marina Del Rey, California, in 1964, where she took her first sailing lesson. One year later, Adams had become quite proficient at sailing and made her first solo trip, sailing from Marina Del Rey to Hawaii on the twenty-five-foot *Sea Sharp*. On this date, Adams completed her most ambitious voyage—a trip from Yokohama, Japan, to San Diego, California, a trip of some 5,600 miles aboard the thirty-one-foot *Sea Sharp II*, making her the first woman to solo across the Pacific Ocean. The voyage took seventy-four days, seventeen hours to complete.

JULY 27–31, 1969

The Dream Is Over

22

T PLUS ELEVEN DAYS

With the astronauts and lunar samples safely back on Earth, there was a fear running through NASA personnel that the lunar samples returned by Apollo 11, and even some of the equipment used by the crew that had made the voyage back, might be the target of thieves; after all, this was the first manned mission to another world, and it could bring in big money in the wrong hands.

As Armstrong, Aldrin, and Collins relaxed in quarantine in the Lunar Receiving Laboratory, NASA headquarters directed its personnel to secure the materials as soon as the quarantine period was over. In a memo, management directed personnel to identify, record, and inventory all recovered items from the flight as soon as quarantine, decontamination, and deactivation activities permitted, no matter how insignificant the item may seem. The memo went on to outline procedures for securing the items:

> All items would be placed in secure storage, under guard if necessary. No removal would be permitted that would deface exterior portions of the spacecraft or portions of the cabin visible through the hatch or windows. No destructive testing would be permitted. Items returned to contractors for testing would be under bond. Preparation for public display should be expedited.[1]

Meanwhile, scientists wasted no time when it came to studying the returned lunar material. Lunar rocks and soil were placed in special sealed glass boxes that had long rubber gloves built into the sides that scientists could use to examine the samples. Only a few days after the material's return to Houston and the LRL, scientists were excited to find that some of the rocks brought back were similar to those found near ancient volcanoes here at home, meaning that at one time, the moon had been an active world with geologic processes much like Earth.

The scientists also discovered significant amounts of feldspar, pyroxene, and olivine, geologic material also found on Earth. This raised investigators' expectations that the Earth and moon actually had similar origins when they were formed. Those men and women taking the first closeup look at the lunar rocks and soil cautioned that these findings were preliminary and that the samples were very limited. Many more samples from several different landing sites would be required to get a full picture of the moon's history and make up.[2]

In another scientific first, NASA announced that some experiments known as the Early Apollo Surface Experiments Package (EASEP) that was left behind on the moon by the Apollo 11 crew were working perfectly. Being that this first moon landing was more of a test flight than a science-oriented mission, only two experiments were included in the package. Later missions would carry seven.

The selection process for determining which experiments should be brought to the moon actually began in March 1963 when NASA, the Jet Propulsion Laboratory, and the Goddard Spaceflight Center held a series of meetings to determine which experiments would provide the maximum return for the least amount of weight and complexity.[3] One of the experiments, a Passive Seismic Experiment Package, would measure meteorite impacts and "moonquakes." The device was so sensitive that when the astronauts turned it on, it registered their footsteps as they walked.

The second experiment was called the Lunar Range Retro-Reflector (LRRR). This little device was a set of silica cubes that could reflect light. Scientists on Earth could shoot laser beams to the LRRR, which would reflect the light back toward the planet. The experiment has provided the exact distance between the Earth and the moon down to about three inches.

The news media also had a treat as the Apollo 11 mission was wrapping up. NASA presented them with a ten-minute screening of the first color pictures and films taken on the moon by Aldrin and Armstrong. Following the showing, the press questioned the actual color produced on the film, since the two astronauts said that the moon's surface was a nondescript gray color, but the photos showed it as purplish. The *New York Times* suggested that women, since they have keener color perception than men, should look at the film before it is fed to the press to make sure that the colors would be adjusted correctly.[4] Weeks later, Armstrong would elaborate on the moon's color saying that he was surprised at the eerie play of light and color.

"At lunar dawn," he told reporters, "the airless moonscape seemed drained of color, but as the sun rose, the moving light was reflected in bright tan.

Surprisingly, though, the actual color of the rock when viewed close-up was dark or charcoal brown. The horizon, close by, was jagged and outlined with knife-edge sharpness against the black abyss."[5]

It was at this time that a former US Navy officer and technical writer for one of the rocket manufacturers of the Apollo program, Bill Kaysing, claimed to have inside knowledge that the lunar landing never happened. Kaysing believed that NASA could not pull off such a feat as landing a man on the moon by the end of the decade so instead, they filmed the entire mission in a film studio. The start of the conspiracy theories that the moon landings were fake, which you still hear to this very day, can be traced back to a book that Kaysing wrote in 1976. Even though every one of his theories has been debunked, over fifty years later, there is still a segment of the population that believes them, even contriving new theories to bolster their own opinion. Polls in the United States show that up to 10 percent of Americans think the landing was a hoax, while in Britain, 12 percent believe the landing was staged.[6] These nonbelievers tend to be very vocal, as Buzz Aldrin found out in 2002 when he was confronted by one denier who got into the astronaut's "space" and called the second man on the moon a liar. The former astronaut had had enough and punched the man in the face.[7]

The United States pulled off another space feat as the historic month of July 1969 was ending. It was one that was almost completely overshadowed by the Apollo 11 mission and didn't receive as much of the public attention, but it was a huge step forward in the exploration of Earth's planetary neighbors. On July 31, the first of two unmanned Mariner spacecraft, Mariner VI, sent the first of thirty-three closeup photographs of Mars from its flyby of the Red Planet.

Scientists at the Jet Propulsion Laboratory in Pasadena, California, were ecstatic with the images, stating that they were the most detailed ever taken of the planet, images that Earth-bound telescopes could never achieve. As the spacecraft flew closer and closer to the planet, taking pictures along the way, each image provided increasing detail of our Martian neighbor. Eventually, Mariner VI came within 2,132 miles of Mars, the closest approach of any spacecraft to date, and returned 75 photographs. Five days later, its twin, Mariner VII, arrived and provided another 126 images, adding to the list of questions scientists had about the planet while laying the groundwork for future unmanned Martian rovers to explore the planet years later.

The euphoria felt by world leaders after the success of Apollo 11 had long-lasting effects, not only for NASA but also in terms of international cooperation to tackle much larger space-exploration ambitions and Earth-related issues. In a letter to President Nixon, NASA administrator Thomas Paine outlined the prospects for international cooperation in space exploration and Earth-bound applications. Paine urged the forerunner of the European Space Agency, the European Space Research Organization (ESRO), to begin serious consideration of new approaches to achieve more participation by European countries, not only in space but also in a new ultra-sophisticated North Atlantic Air Traffic Control System to improve passenger air safety and build upon and improve the existing global meteorological satellite system for weather prediction.

In a letter dated December 15, 1969, addressed to the chairman of the Senate Committee on Aeronautical and Space Sciences, Senator Clinton P. Anderson, Paine summarized his efforts to gain more international cooperation in space and, in particular, requested that other countries consider "sponsor[ing] their own industrial participation in the NASA conference on space shuttle concepts [in October]."[8] He also announced a new agreement had been reached with Japan to purchase space technology from the United States.[9]

Paine was also keen on restarting dialogue with the Russians concerning joint cooperation in space, in particular, the director proposed that the Soviet Union be a partner in the upcoming unmanned Viking missions to Mars. He also offered the Russians use of the lunar reflector left on the moon by Apollo 11 and invited proposals by Soviet scientists for lunar sample analysis. Finally, the director left the door open for further talks on other possible areas of cooperation stating that the United States was ready to "meet anytime, anyplace, to consider any possibilities for cooperation or coordination between [the two countries]."[10] No substantial response to the outreach was given by the Kremlin.

As for the Apollo program, despite significant budget cuts by the federal government and the waning of public interest after the first moon landing (the mood of the country had turned to one of, "been there, done that"), NASA announced plans and landing sites for the next nine Apollo missions:

Apollo 12	November 1969	Oceanus Procellarum
Apollo 13	March 1970	Fra Mauro
Apollo 14	July 1970	Crater Censorinus
Apollo 15	November 1970	Littrow Volcanic Area
Apollo 16	April 1971	Crater Tycho
Apollo 17	September 1971	Marius Hills
Apollo 18	February 1972	Schroter's Valley
Apollo 19	July 1972	Hyginus Rille
Apollo 20	December 1972	Crater Copernicus

But it didn't take long for reality to set in. After humankind's crowning achievement, breaking away from the bonds that hold us to Earth to explore another world, with public interest waning and Congress continuing to cut the space agency's budget, it became too much for the American space program to bear. In the end, the final three Apollo missions would never fly, and any dreams of creating permanent manned lunar bases, even landing humans on Mars, were scrapped. NASA began to look at more Earthly applications for spaceflight.

On November 14, the second trip to the moon began with the launch of Apollo 12. For the most part, Americans were space weary, and the event was hardly noticed, although the flight began with an ominous message as to how dangerous space flight really was.

It was a rainy morning for the launch. A cold front had moved through Florida the night before the launch producing numerous thunderstorms, but by launch day, the weather had markedly improved although some light showers persisted. Apollo 12 was given the go for launch. In the crowd watching were President Nixon and First Lady Pat Nixon.

The countdown continued without a hitch and at 11:23 a.m., the five large Saturn V engines ignited, shaking the Earth and producing its telltale explosive roar. The stack of machinery pitched over slightly, as programmed, to avoid hitting the launch tower, and seconds after it had cleared the tower, a loud crackle was heard in the communications loop in the launch firing room while inside the capsule *Intrepid*, warning lights and buzzers began to blare.

"What the hell was that?" command module pilot Richard Gordon shouted. "I lost a whole bunch of stuff. I don't know . . ."

Mission commander Pete Conrad jumps in. "Okay. We just lost the platform, gang. I don't know what happened here. We had everything in the world drop out."[11]

What had happened was that a lightning bolt struck the Saturn V only forty seconds after launch and all their electric power generation had failed. As the crew and mission control tried to figure out what was happening, only twenty seconds later, a second lightning bolt struck causing their inertial guidance system to spin wildly. The spacecraft had lost its bearing during one of the most dangerous moments of the flight. It didn't know how high or fast it was flying, and without such information, the Saturn could crash or explode.

In Mission Control, engineer John Aaron realized that the electrical issues were something that he had previously experienced during factory testing of the capsule. He knew the answer—turn the Signal Conditioning Equipment to auxiliary. The command was given: "SCE to auxiliary."

"What the hell is that?" Conrad questioned.[12]

The correct actions were calmly relayed to the astronauts and lunar module pilot Alan Bean radioed back, "Everything looks good."

A tragedy was averted.

It was about this time that the team in Huntsville at the Marshall Spaceflight Center, the same team that had designed the Saturn V rocket that took men to the moon, was given a new mission: to create an orbital workshop—a space station—that would be sent into Earth orbit by 1972. The workshop would be launched into space using a leftover Saturn V, sending the third stage, which would be converted into the actual space station itself complete with living quarters and scientific experiment stations, into Earth orbit. After deployment, a three-man crew would be launched aboard an Apollo command module atop a smaller version of the Saturn V rocket, the Saturn IB, to dock with the space station and live in Earth orbit for a period of up to fifty-nine days in space. The project would be called Skylab.

A number of experiments were already underway for sending humans into space for extended periods of time—how to make life comfortable for them, what they would eat, even how to take a shower in space. The latter was intriguing: how to take a shower in the weightless environment of space. How could randomly scattered water droplets be corralled to make that happen?

Experiments were conducted by several organizations and universities on this very subject in an attempt to determine if zero-gravity showers were possible. In one experiment conducted by the Marshall Spaceflight Center, an experimental shower and test subject, Jack Slight, were placed inside a KC-135 jet, better known as the "Vomit Comet," to try it out. The jet would fly in a parabola where at the top of its arc, passengers onboard would be weightless for up to twenty-five seconds. C. C. Johnson with Marshall summarized the results of the test in a letter to NASA headquarters.

"MSC has some excellent films of Jack Slight showering in the KC-135 at zero-gravity. The motion pictures of Jack showering are quite revealing—not of Jack, the action of water at zero-gravity. . . . The interesting point is that the water strikes Jack, bounces off in droplets, but then recollects as jelly-like globs on various parts of his body. He can brush the water away but it will soon reattach elsewhere."[13]

As the Apollo program was winding down and the US space program was looking toward a more permanent outpost in Earth orbit, the Soviet Union had already shifted gears and was in the process of beating the Americans in space station development, but it would be a difficult road to get from that first Russian space station to the International Space Station.

Two years after Apollo 11, the Russians orbited Salyut 1, the world's first space station. This first tentative step toward extended living in space was plagued with failures: The first crew to the station aboard Soyuz 10 was unable to dock with the station due to docking mechanism failures. The second crew managed to dock and live on Salyut 1 for three weeks, but they perished on their return to Earth when their capsule experienced a pressurization problem. This was followed by three next-generation stations that failed to either reach orbit or broke up in space before crews could reach them.

In the United States, the afterglow of the first lunar landing was burning out. An editorial in the July 28 issue of the *Nation* magazine asked, "What Price Moon Dust?"

> Had there been no Soviet Satellite in 1957 Americans would not by now be reaching for the moon. . . . Much has quite rightly been said about the irony of spending billions getting to the moon while the mass of humanity at home lives in a stew of exploding population, poverty, and pollution. . . . The time of decision is here and the euphoria of the moment—however understandable it may be—must not be allowed to obscure our judgement.[14]

The shiny object that was Apollo was gone.

23

THERE'S A MANIAC LOOSE OUT THERE!

The crime rate in the United States had been on a steady rise since 1960 and was continuing to skyrocket in 1969. In ten years, violent crimes had gone up 126 percent,[1] with the number of murders jumping from 9,110 in 1960 to 14,760 in 1969.[2] Of those murders in 1969, an exorbitant number of them were linked to serial killers, a term the FBI attached to any killing in which there are two or more victims of a crime carried out by the same person (or persons). On July 12, one of those killers, the Cape Cod Vampire or Cape Cod Cannibal, Tony Costa, confessed to the brutal murder and mutilation of two young women in Truro, Massachusetts.

Tony Costa began showing signs of sociopathy when he was in his late teens. Twice in November 1961 he made nighttime invasions into the bedroom of a sixteen-year-old girl in her family's apartment. The first time, he stood over her bed and stared at her and didn't flee until the girl woke up and started screaming. The second time, he grabbed the girl and attempted to drag her downstairs into the building's basement. A neighbor intervened and saved the girl's life.

Over the years, Costa served a few short stints of time in jail for burglary before marrying a young woman who was fourteen. The couple had three children but soon, he began using drugs and exhibiting bizarre and irresponsible behavior. In 1966, the marriage was on the rocks, and Costa decided to head to California.

As was the case in the late 1960s, Truro was a transient area where hippies gathered and communed. It was a common occurrence for teens to arrive in town looking to live the hippie lifestyle then disappear without notice and sometimes without a trace. When Costa left for California, he picked up two women hitchhikers, Bonnie Williams and Diane Federoff, and offered to drop them off in Pennsylvania. It was on his way. The two women were never seen again and are believed to have been Costa's first victims.

In 1968, Costa was heading back to Massachusetts from California. He had a new girlfriend, Barbara Spaulding, who disappeared the same day Costa was leaving on his return trip.

In May, Costa was back in Truro where he met Sydney Monzon. That month, Monzon told her sister Linda that she "wanted a new life, one filled with adventure and exploration." She had only graduated from high school the year before but was looking to break away from the ordinary.[3]

Sydney was described as having a radiant smile and being able to make friends easily. She was attracted to intellectuals, someone like Tony Costa, and once the two met, they were often seen together riding bicycles through town. She called Costa, "Sire."

On May 24, 1968, Sydney's sister Linda saw Sydney at the top of a hill standing next to a car. Sydney shouted to her sister and asked if she could come up to the car. "I can't, Sis," Linda Monzon shouted back. "I'm already late." Sydney had a troubled look on her face, but her sister waved it off as being nothing to worry about and continued on her way. Sydney climbed into the car. It belonged to Costa.

The young woman asked Costa where they were heading, and he replied that they were going to get some pills. "And then I'll take you to a place where a thousand tiny Tinkerbells will descend upon you and carry you to fantasy's domain."[4]

The pair headed off in the night in Costa's Oldsmobile to a local drugstore. Sydney, who was driving the car, dropped off Costa to "pillage" the store. The plan was that he would take the pills to a spot near an old cemetery where the couple often met up to get stoned. He would meet up with her there and they would split up the pills.

When they met at the cemetery, it was pouring rain. This is when Costa's split personality made an appearance. He named his second personality Cory. "Don't do it, Cory," he told his alter ego. "She doesn't deserve to die."[5]

Taking out a knife that he had stashed in one of the bags he used to carry the drugs in, Costa began chasing the young woman. When he finally caught up with her, he plunged the knife into her repeatedly.

The next day, Sydney's sister Linda confronted Costa, who denied knowing where her sister was. Several days later, Linda told her parents that Sydney was missing, and they filed a missing person's report with the police. As mentioned, a missing teen was nothing new in Truro. The previous year, nineteen teenagers had gone missing, but no one, not even their parents seemed concerned. It was a town of transients, after all.

By September, as the search continued, Costa had yet another girlfriend, Susan Perry, who vanished mysteriously as well. Costa told people who asked about her whereabouts that she had "gone to Mexico."[6]

In January 1969, Patricia Walsh and Mary Anne Wysocki also vanished without a trace. It wasn't until February that police finally had a lead, finding the women's VW beetle near the town's old cemetery.

Officers searching the woods made a gruesome discovery—a bag containing dismembered body parts. It was the remains of Susan Perry. Continuing the search, they discovered the mutilated body of Patricia Walsh, and beneath her, Sydney Monzon. Some of the bodies had bite marks on them, while others could only be identified by the jewelry they wore, such as the wedding ring one victim's mother had given her.

Costa was arrested and charged with the murders. The prosecution described him as being the most "vicious killer since Jack the Ripper." While awaiting trial in July 1969, Costa spoke with the district attorney and maintained his innocence giving the impression that he had no idea that he was the killer police were looking for. At one point he asked the attorney what he was doing about the murders. "There's a maniac loose out there!" he shouted.[7]

The crime scene became a carnival side show with tourists arriving with shovels and picnic lunches to "help" forensics teams dig. One man began selling bags of sand from the grave site as souvenirs for fifty cents a bag.[8]

The story finally comes full circle where it started. Costa eventually confessed and was sentenced to life in prison for the murders of Patricia Walsh and Mary Anne Wysocki. It was never proven how many women fell victim to Costa's brutal rampages, and the world will never know. Costa died by suicide in prison in 1974.

Out in California, another true-life murder mystery was playing out. It started with the single murder of a young couple in December 1968, but in July 1969, local, state, and federal law enforcement officials realized that they were dealing with more than just a simple murderer running loose. They were dealing with a deranged serial killer.

The killer in question targeted and shot random young lovers that he came across on the deserted backroads of Vallejo, a San Francisco suburb. Soon, he began sending cryptic messages to the police about the murders, taunting them, daring them to try and find him. It was a mystery that would grip the state in fear for years to come, perpetrated by one of the most infamous serial killers in US history, the Zodiac Killer.

The Zodiac first came into the public spotlight on December 20, 1968. Seventeen-year-old David Faraday and his sixteen-year-old girlfriend, Betty Lou Jenkins, were out on a date in Faraday's station wagon. This was the first time that Betty Lou's parents had allowed her to go out on a date with a boy.

The couple were parked in a gravel parking lot at Lake Herman when at about 11:15 p.m., the couple was approached by an unknown man. The man pulled out a gun and fired shots forcing the couple out of the car.

Betty Lou was the first to step out of the vehicle with David exiting close behind. The man aimed the pistol at David and shot him point blank in the head. Betty Lou tried to run but was shot five times in the back. No motive for the killing, or a suspect, was ever found. At the time, it seemed like a random, senseless, and tragic killing, but in the following months, law enforcement in the Bay Area would soon realize that this was the work of a deranged serial killer.

The second murder came in the early morning hours of July 5, 1969. One of the victims, twenty-two-year-old Darlene Ferrin, was described by friends as a person who had magnetic charm and was instantly liked by all who met her. She was a waitress at a local restaurant, Terry's, and she and her husband, Dean, were the parents of an infant.

Despite being married, Darlene was known to have been dating nineteen-year-old Mike Mageau. On that warm summer night in 1969, the couple made plans to see a movie in San Francisco, but due to unexpected delays, they didn't meet up until around 11:30 p.m. Since it was such a late hour, the couple decided to change their plans and get something to eat instead.[9]

Ferrin drove over to pick Mageau up from his home in her brown Corvair. On the way to the restaurant, she told the young man that she needed to talk to him about something. She turned her car into an empty parking lot, turned off the lights and engine but left the radio playing softly in the background.

As the couple sat talking in the dark parking lot, a car drove in with a loud group of teenagers shouting and shooting off some firecrackers before driving away. A few moments later, a second car pulled into the parking lot, approximately eight feet behind Ferrin's car and turned off its lights.

Only a few minutes later, the car drove off. The couple continued their conversation, only to be interrupted yet again by the return of the same mystery vehicle. This time, it parked more toward the passenger's side of Ferrin's car and left its headlights on. The positioning of the car made it difficult for Ferrin and Mageau to see that the man driving the mystery vehicle had stepped out of his car and was approaching them. The man wielded a high-powered flashlight much like a police officer would use, so naturally, the couple assumed that it was a police officer checking up on them.

As the man approached the opened passenger side window, the couple began pulling out their identification. The mystery man pulled out a gun and fired five shots into the car. Believing he had killed the couple, the gunman started back to his car when he heard Mageau scream out in pain. The gunman walked back and shot the couple two more times before driving off.[10]

Miraculously, Mageau was still alive and managed to open the passenger door and fall to the ground. A short time later, another car pulled into the parking lot and discovered the gruesome scene. The police arrived soon after and transported the couple to the hospital. Darlene was pronounced dead on arrival, but Mike survived and gave police a description of the killer. However, even with the young man's description and fingerprints left behind by the murderer, no one was ever brought to justice for the crime.

Within an hour of the shooting, a man contacted the Vallejo police department to report the crime and to announce that he was the killer. He also proclaimed that he was the killer of Faraday and Jenkins the previous year. In the upcoming weeks, the killer sent a letter to three newspapers—the *San Francisco Examiner*, *San Francisco Chronicle*, and the *Vallejo Times-Herald*. The letters began, "Dear Editor: I am the killer of the two teenagers last Christmas at Lake Herman." The letter went on to claim responsibility for the second killing as well and provided details of the murders that only the true killer would have known. The letter ended with a cryptic message that became the killer's trademark—a circle with a cross in it, and a cipher that allegedly contained his identity. Police called him the Cipher Killer. The killer called himself Zodiac.

The Zodiac was linked to at least five more murders over the following years and was a suspect in several more, all of which were followed with the same taunting letters and phone calls. The last murder that was linked to the Zodiac occurred in 1974, after which all of the letters ceased. In 2018, over fifty years later, a group of forty investigators acting on their own claimed to have identified the killer, but the suspect had died that same year. In 2021, the FBI announced that the investigators' claim was incorrect, and the case of the Zodiac Killer remains open.[11]

In the same month that the Zodiac Killer began his reign of terror, another infamous killer also began his own spree. This time, it wasn't a single man committing the murders but a group of young people who were caught up in a cult that was led by one man, Charles Manson.

Manson's criminal career began when he was nine years old. At the time, he was being raised by an aunt and uncle in West Virginia because his mother was in prison for armed robbery. Manson spent his early years in and out of reformatories and prison for petty theft, auto burglaries, and armed robberies.

After being released from Terminal Island Prison in 1967, Manson moved to Berkeley, California, where a full-blown cultural revolution was taking place. Berkeley had become the epicenter of the free speech and anti-war movement, something Manson had never seen before since he had been in prison most of his life. For Manson, the world had changed.[12]

Manson fit right into this culture and began his own communal "family," one in which he professed eccentric religious ideas and predicted an apocalyptic race war that would decimate the general population leaving the Family as the world's dominant power. His cultivation of the Family began one day as he was living on whatever money he could earn from panhandling at the gates of the University of California. It was here that he met Mary Brunner, a university librarian. Manson told one of the Family's future members, Danny DeCarlo, about the encounter.

"She was out walking the dog," DeCarlo recalled Manson saying. "High button blouse. Nose stuck up in the air, walking her poodle. Charlie's fresh out of the joint and along he comes talking his bullshit."[13]

With his nefarious charm and hypnotic eyes Manson was able to manipulate Brunner, promising a life of family, enlightenment, and home.

"So one thing led to another," DeCarlo continued. "He moved in with her. Then he comes across this other girl. 'No, there will be no other girls moving in with me!' Mary says. After the girl moved in, two more came along. 'I'll accept one other girl,' Mary says, 'but never three.' Four, five, all the way up eighteen [arrived]."[14]

For a time, the Family lived in the home of the Beach Boys' drummer, Dennis Wilson, but when they were evicted, the members of the cult found an old ranch to move to—Spahn Ranch. The ranch was owned by eighty-one-year-old George Spahn. At one time, the ranch and its old 1800s-style buildings were used for western movie and TV show productions in the 1950s. Spahn, who was blind, agreed to let the Family stay at the ranch in the long-since-abandoned shacks as long as they helped with the chores.

According to author James Buddy Day, "The ranch really isolated the women. There were no books, clocks, or calendars. They became increasingly reliant on each other, which enabled their eventual feelings of paranoia and fear, all culminating in the murders."[15]

It was here at the ranch that the first Manson Family murder occurred, and it set off a chain of violent and gruesome killings. It was the murder of thirty-four-year-old Gary Hinman.

Beausoleil was an up-and-coming pianist who had one time played Carnegie Hall. In an interview with *People* magazine, Beausoleil's cousin, Charlotte Hood, described the young man as a "lost artistic soul. A talented piano player with a collection of college degrees who flittered from job to job."[16] Like others, Beausoleil fell under the spell of Manson.

In July 1969, Beausoleil bought drugs from Gary Hinman. Hinman was thirty-four years old and had a degree in chemistry. At the time, Hinman was a student at UCLA and was working on his PhD in sociology. He was introduced to Manson and the Family in 1968. From there the story gets murky.

One story says that Manson had heard, erroneously, that Hinman had inherited a great deal of money. Another says that Hinman had been invited to join the cult, but he declined due to being a devout Buddhist. The third story claims that Beausoleil had bought drugs—a thousand mescaline tablets—from Hinman that he believed were of inferior quality. Whatever the story, it definitely all boiled down to money and the ending was the same.

On July 25, Mary Brunner, Bobby Beausoleil, and another of the Manson women, Susan Atkins, went to Hinman's house brandishing a gun demanding money. Hinman pleaded with the three, telling them he didn't have any, even showing them that he only had $50 in his pocket.

Manson was called and was informed that Hinman was not forthcoming with the cash. Not long after, the cult leader showed up at Hinman's home in person brandishing a Samurai sword. Manson told Beausoleil that he needed to be shown "how to be a man" and cut Hinman's face with the sword from his left ear down the side of his face. Brunner and Atkins attempted to stitch the wound back together with dental floss.

On July 27, Beausoleil returned and stabbed Hinman twice in the chest. The young man clung to life for several hours and finally died after the two women took turns holding a pillow over his face to quicken his death. When Hinman finally died, the three Manson followers began what would become a trademark of Manson Family murders, writing "Political Piggy" in the victim's blood on the wall. Hinman became the first of the Manson Family murders.

Beausoleil was found by police ten days later sleeping in one of Hinman's cars alongside a highway. The murder weapon was hidden in a wheel well. He was convicted of first-degree murder and sentenced to death, but the sentence was later commuted to life in prison.

Susan Atkins was later arrested and found guilty in connection to one of the most famous Manson murders, the Tate-LaBianca murders, the following month, during which pregnant actor Sharon Tate and several others were killed in what later became known as the Helter Skelter Murders. Brunner, who was the mother of Manson's child, was sentenced to twenty years in prison for the same murder but was paroled in 1977. She disappeared from the public eye immediately after her release.

24

THE SEX EXPLOSION

The phrase *free love* was bandied about quite a lot in the late 1960s. It was a phrase that meant different things to different people. In general terms, it meant that people should love one another not physically but by showing consideration, kindness, and empathy to others.

For some, "free love" meant not being tied down with the legalities of marriage. A couple could live together, have sex, have children, but there were no binds tying them together. For the most part, these couples were monogamists and were devoted to their partner. For others, it meant the freedom to physical—consensual or not—lovemaking with multiple partners. Of course, the parents of those who professed free love only saw it one way: it meant debauchery, depravity, and lasciviousness.

The losers in the free-love movement were women. Some women did enjoy the concept of the *sexual revolution*, but the majority suffered because of it. Men dominated the hippie movement, and many of them believed that free love meant free sex. In a 1967 essay, Chicago-based feminist Evelyn Goldfield wrote, "I think that there was a general feeling that the whole idea of free love was a very attractive idea to men because it meant love without responsibility."

Many women found that the sexual revolution of the late 1960s did not equate to equal opportunities and status in society. "Women were used as inducement to get new members into a commune," Dr. David Smith recalls of the time period. "If you joined, you got to have sex with the girls."[1]

The rates of sexually transmitted disease in San Francisco alone rose 150 percent in one year, as did the number of back-alley abortions which often ended with tragic results. From all of this, women began to rise up and turn the tables on the sexual revolution as they began seeing themselves as equals with men if not taking the lead. Women became empowered to stand up for their rights and autonomy over their own bodies and to decide if and when they would have children. Those that had children began raising them to be assertive and determined in their beliefs and the direction in which they wanted to lead their lives. While there is still much work to be done on this front, women's rights have come a long way from the free-love movement of the late 1960s.

With the sexual revolution being an undercurrent of society in July 1969, it was becoming clear that sex was becoming more and more mainstream in the arts and pop culture. This is evidenced in the July 11, 1969, issue of *Time* magazine, which led with a cover story titled "The Sex Explosion." The cover showcased a nude man and woman hidden behind a giant leaf with a zipper embedded in it that was pulled halfway down. The story behind the image was blunt and to the point.

> The issue is as old as the fig leaf, as new as tomorrow's nude-theater opening.
> An erotic renaissance (or rot, as some would have it) is upon the land. Owing to
> a growing climate of permissiveness—and the Pill—Americans today have more
> sexual freedom than any previous generation. Whatever changes have occurred
> in sex as behavior, the most spectacular are evident in sex as a spectator sport.[2]

Sex as a spectator sport. Quite the summation, but in the summer of 1969, that is what it seemed like as Broadway theatrical productions and mainstream movies, books, even television were beginning to explore how far they could go with this newfound sexual openness. Comedian Russell Baker commented on the trend in a series of photographs published in *Life* Magazine. "The nudity crowd tells you 'the human body is beautiful'. . . . Except for a few oddities, like Tarzan or Raquel Welch, the human body is one of the ugliest bodies around. This is because it's not covered with hair, scales, hide, or feathers. Did you ever see a snake with a potbelly? An antelope with varicose veins?"[3]

One of the first big-screen attempts at pushing the barriers was the movie *Midnight Cowboy* that was released on May 25, 1969. The movie starred Jon Voight as Joe Buck and costarred Dustin Hoffman as Enrico "Ratso" Rizzo.

Buck is a young man from Texas who quits his job as a dishwasher and heads to New York City where he believes he can sell his sexual "services" to the rich, sex-starved women of the city. The naive Texan's first "job" was with a woman whom he brings up to his apartment, only to find that she is a call girl herself. He ends up paying her instead.

From there, the movie follows Buck through a series of mishaps that include steamy scenes of homosexuality. The entertainment world was shocked at the movie's content and depiction of sexuality, so much so that it was given an "X" rating by the Motion Picture Association of America (MPAA), meaning that only adults over the age of eighteen could pay to see it. And they did.

The $3.6 million budget film raked in over $44 million in box office sales, making it the number 1 motion picture in July 1969. The movie garnered six BAFTA (British Academy of Film and Television Arts) awards, seven Golden Globes, and three Academy Awards, one of which was the Best Picture award, making it the first—and only—X-rated movie to win the Oscar.

Eventually, the movie's rating was reduced to "R" (restricted), and it was added to the American Film Institute's Top 100 Movies list.

X-rated movies were nothing new, having first appeared almost as soon as the motion picture camera was invented. But these movies were normally poorly filmed shorts with the most basic of camera equipment that produced grainy images and were only shown in the seediest movie houses on the outskirts of towns. There were a few, however, that transcended that stereotype and were actual full-length movies with thin plots (if any at all), which gave rise to *Time*'s analysis that sex was now a "spectator sport." One of those movies, while not of Academy Award–winning quality like *Midnight Cowboy*, was released in July 1969 and actually made a killing in mainstream movie theaters—*The Stewardesses*.

The storyline for *The Stewardesses* focuses on a single day in the lives of several, well, stewardesses, who are based out of Los Angeles, and their experiences with drugs and sex. Lots of sex. The film was made with a budget of only $100,000, but by the end of July 1969, it had brought in over $13.5 million ranking it the fourteenth-highest grossing film of 1969, placing it just behind the likes of *Midnight Cowboy*, *True Grit*, and *Hello, Dolly!*[4]

Even with the successes of those two movies, one of which was deemed as art, sexually explicit motion pictures were not the norm on the big screen. As mentioned, the bulk of the films produced were straight-up pornography and were relegated to theaters on the bad sides of towns. In New York City, the once-thriving theater district of Forty-second Street, with such grand theaters as the Liberty, Rialto, and New Amsterdam had descended fully into a strip of porn theaters, go-go bars, adult bookstores, and peep shows.[5] Revitalizing the street and bringing it back to its former glory would take decades, due in part to the fact that each of the adult theaters lining the street was bringing in gross incomes of up to $100,000 weekly.[6] A difficult sum of money for store owners to give up.

One theater, at the corner of Second Avenue and Twelfth Street was one of those that had succumbed to becoming a pornographic cinema house. In 1969, it was transformed back into a theatrical playhouse and would be where a sex-centric play would be introduced to the world—*Oh! Calcutta!*

Nudity was not new in the theater world, but until *Oh! Calcutta!* opened, Broadway productions only had a brief topless scene or two, if that much. In 1965, the actor playing the part of Jean-Paul Marat briefly flashed the audience during the run of *Marat / Sade*. But *Oh! Calcutta!* would take theater to a different place.

The play was conceived by English theater critic Kenneth Tynan, who renovated the old Eden Theater so that it could be used as the home of his new production. The play would be a series of musical numbers and comedy sketches about sex and sexual mores. The sketches for the play were solicited by Tynan from the likes of actor Sam Shepard, cartoonist Jules Feiffer, and Beatle John Lennon.

Actress and choreographer Margo Sappington was approached by director Michael Bennett to try out for a role in the play and to be his assistant choreographer. "I was his assistant in *Promises, Promises,*" she recalls. "He came to my dressing room between shows. He said, 'I need you to come and help me. And oh, by the way, you have to be naked.' When he decided he wasn't going to do the show, the producers asked for a recommendation. He said, 'Margo can do it.'"[7]

Actress Raina Barrett read the casting call that mentioned the play would include nudity. "Well, we don't know how much nudity," she thought to herself. "I'll consider it."

As it turns out, there was a lot. The play did not need a major costume budget. Most of the two hours had the actors completely nude. Following its opening on June 17, 1969, *New York Times* theater critic Clive Barnes called the show "witless" and commented that "this is the kind of show to give pornography a bad name."[8]

While that may have been true, by the middle of July, *Oh! Calcutta!* was one of the hottest theatrical tickets in the city proving in a way that sex sells. On July 17, it was announced that ticket prices would rise from $7.50 to $25, an exorbitant amount of money for the time, which made it one of the highest ticket prices in the city.

But it wasn't only in motion pictures and theater where the sexual revolution was making inroads into the mainstream. The publishing world was also embracing this titillating world, and it wasn't in the adult-oriented men's magazines of the day. It was in a world where women could lose themselves in passion—the romance novel. The most popular author of the genre was Jacqueline Susann, with her incredibly successful novel *The Love Machine*.

Erotic romance has been around since humans first learned to write, but actually putting steamy romance into novel form did not occur until the eighteenth century with the publication of John Cleland's *Memoirs of a Woman of Pleasure*, better known as *Fanny Hill*, in 1748. The novel wasn't explicit in nature. By using literary devices, Cleland used euphemisms to describe the act of love making. The book is one of the most popular—and most banned—books in history. Even the US Supreme Court weighed in on whether or not the book was obscene in the case *A Book Named "John Cleland's Memoirs of a Woman of Pleasure" v. Attorney General of Com. of Mass.*, in which the higher court ruled that a previous court ruling on the subject proved that the material met the legal standard of being "unqualifiedly worthless," thus to be pornographic.[9]

The *Love Machine* was released in May 1969 and was Susann's second novel. Her first, *Valley of the Dolls*, was panned by many critics, but that didn't stop its female readers from being swept away in the storyline and imagery. That book sold over thirty-one million copies.

The publisher summarized the premise of her latest book: "In a time when steak, vodka, and Benzedrine were the three main staples of a healthy diet and when high-powered executives called each other baby, network tycoons had a name for the TV set: *The Love Machine*. But to supermodel Amanda, socialite Judith, and journalist Maggie, the love machine meant something else."[10]

You can see where this is heading. The *New York Times* called the book a "very long, absolutely delicious gossip column"[11] while *Times* freelance writer and romance novelist Nora Ephron, said of the book, "Most women, I think, do not want to read hard-core pornography. They do not even want to read anything terribly technical about the sex act. What they want to read about is lust, and Jacqueline Susann gave it to them. . . . Hot lust. Quivering lust . . . [where] everyone is wanting everyone else."[12]

Ephron was right. On July 6, *The Love Machine* went to number 1 on the *New York Times* Bestseller List selling an astonishing three hundred fifty thousand hardback copies and an even more astonishing eight million in paperback.

And then there was the music scene, which tended to get into more trouble than many other artforms of the late 1960s for their lewd lyrics, sensual sound effects, and erotic album covers. Of the latter, the most provocative was a debut self-titled album called *Blind Faith* by the band of the same name that was released in the United States on July 21. The band consisted of rock legends Steve Winwood, Ginger Baker, Eric Clapton, and Rick Grech. The album was hastily recorded following a sellout tour of the United States.

In America, the album cover was simply a photo of the band gathered around a drum kit. When the album was released in the United Kingdom the following month, the vinyl sported a different cover—a photo of a topless eleven-year-old girl. As you can imagine, record shop owners refused to sell the album until the cover was replaced. It was, but album sales still plummeted. It was the first—and last—album released by Blind Faith.

Serge Gainsbourg was one of the most popular, and some argue the greatest, French singer-songwriters of the 1960s and 1970s. Born in Paris in 1928 as Lucien Ginsburg, the young man attended the Montmartre Art School where he studied painting. He also had a love of music, especially Chopin, Gershwin, and Cole Porter. His father, Joseph, taught him how to play the piano, and at the age of thirteen as the Germans began their occupation of France, his family was forced to wear a Jewish Star of David on their clothes with the word *Jew* sewn into it. He often told people that the experience scarred him for life.

In 1954, Ginsburg began writing his own songs under the name Lucien Gainsbourg. After two unsuccessful marriages, Gainsbourg met actress, singer, and sex-symbol, Brigitte Bardot, while the two were appearing as guests on a television show. Although Bardot was married, the two began a secret love affair in 1968. During that time, the couple recorded several of Gainsbourg's songs including the 1969 "Je t'aime . . . Mon non plus" (I love you . . . Me neither).

The tune sounded like a beautiful French love song with lyrics that were "a clever interplay of cynicism and sentiment," that is until just over two minutes into the record when suddenly, passionate moans and groans fill the air. Rumor had it that the couple were having sex while recording the song.

Bardot's husband heard the recording and as you would guess, became enraged and demanded that the recording company cancel releasing the single. The company relented but Gainsbourg wasn't going to let it go. Bardot eventually returned to her husband and Gainsbourg met twenty-two-year-old Jane Birkin. In July 1969, Birkin and Gainsbourg re-recorded the song, complete with the heavy petting, only this time, the single was released.

Immediately the BBC and several radio broadcasters from other European countries banned the record. Even the Vatican declared the single obscene. Despite the condemnation, or maybe because of it, the record went on to sell over four million copies. Once again proving that sex sells.

25

IF SHE COULD ONLY COOK AS WELL AS THE HONEYWELL CAN COMPUTE

The farthest thing from anyone's mind as July 1969 was ending was Christmas. That is, everyone except for the marketing magicians who were prepping for the blitz of commercial and print advertising that would be hitting mailboxes, TV screens, and radio airwaves the moment the Macy's Thanksgiving Day Parade had ended, ushering in the festive time of year. If a company was to produce and get its catalogs in the mail in plenty of time for the holiday season, it needed to have them completed and printed by the end of July.

Each holiday season, department stores and catalog retailers like Montgomery Ward, Spiegel, and Sears published their highly popular and anticipated holiday catalogs that would be sent to millions of households across the United States. There was another company that also had a catalog, but this one had a limited audience. It catered to the wealthy, offering only the finest (and often most expensive) merchandise and gift items that recipients "never knew they needed"—the Nieman Marcus catalog.

The Nieman Marcus catalog actually started in 1915 as a simple Christmas card sent to shoppers, inviting them into the retailer's store in Dallas, Texas. The first actual catalog was published in 1925 and was only sixteen pages long.

Since that time, the Nieman Marcus Christmas catalog has promised luxurious home furnishings, the latest elegant fashions, and the unusual. For the discerning shopper in 1969, the catalog offered two live gerbils (which they touted as being cleaner than guinea pigs), one male and one female, with "fantastic family potential," complete with cage, food, food dispenser, water tube, and care booklet for only $35. You could also get that special someone a Baccarat crystal server that would be filled each month for a year with a tin of the finest giant-grain Beluga caviar from Iran for only $600. And for that extra special gift, how about a computer? But not just any computer—a Kitchen Computer! That's right. The Honeywell corporation had taken their computer technology and made it accessible to the average home. Well, sort of.

If she can only cook as well as Honeywell can compute.

Her souffles are supreme, her meal planning a challenge? She's what the Honeywell people had in mind when they devised our Kitchen Computer. She'll learn to program it with a cross-reference to her favorite recipes by N-M's own Helen Corbitt. Then by simply pushing a few buttons obtain a complete menu organized around the entrée. And if she pales at reckoning her lunch tab, she can program it to balance the family checkbook. **84A** 10,600.00 complete with two week programming course **848 Fed with Corbitt data:** the original Helen Corbitt cookbook with over 1,000 recipes 5.00 (.75) **84C** Her Potluck, 375 of our famed Zodiac restaurant's best kept secret recipes 3.95 (.75) Epicure **84D Her tabard apron,** one-size, ours alone by Garden House in multi-pastel provincial cotton 28.00 (.90) Trophy Room

Nieman Marcus added the Kitchen Computer to their 1969 Christmas catalog in July. A handy device that literally took a software programmer to operate. Not a single computer was sold. *Computer History Archive*

The ad that would appear in the holiday catalog that year showed a woman dressed in a brightly decorated Chanel apron adorned with butterflies and flowers and with sandals on her feet admiring her new Honeywell Kitchen Computer, which looked like something straight out of *The Jetsons*. The unit, which weighed over a hundred pounds, was a version of the company's H316 business computer mounted into an ultra-sleek and futuristic black, white, and orange base. The ad from the male-centric catalog started with the headline, "If She Could Only Cook as Well as the Honeywell Can Compute."[1]

Nieman Marcus suggested that a housewife slaves away in the kitchen all day to cook up souffles that are sublime, but then asked, "Is her meal planning challenging?" To which they provided the answer—the Honeywell Kitchen Computer. With the Kitchen Computer, the housewife could cross-reference recipes created by chef Helen Corbitt "by simply pushing a few buttons [to] obtain a complete menu organized around the entrée."

Sounds too good to be true for the time, right? A menu with the push of a button? As it turned out, it would take pushing more than just a few buttons. Reading further down in the fine print, you realize that she would have to "learn to program" this beast. The user would have to use the buttons on the computer to create a series of zeroes and ones, the basis for all computer programming. Push the button in, the computer recognizes it as a one. Release the button, it is recognized as a zero.

But wait. There's more. Once the zeroes and ones have been entered into the computer, hopefully in the proper order, a paper tape would be printed with a series of holes in it that correspond to those zeroes and ones just entered. Think of the paper tape as the compact disc or thumb drive storage of the day, only one with extremely limited capacity (barely any) and very fragile. Any number of things could tear the tape or rip the holes in it.

Hold on! There's even more. After generating the paper tape, the user would then have to take that tape and feed it into a special reader that would transfer those zeroes and ones into its "databank" and voila! Now you can get back to cooking. And all this techno-wizardry could be yours for only $10,600. The machine came complete with a two-week programming course. Seriously.

The computer did, in fact, exist, but as with many Nieman Marcus catalogs of the past, it was more of a publicity stunt for something that really wasn't attainable for the average person. Not a single kitchen computer was sold, but it got readers thinking about the possibilities—a feature that the catalogs became known for. For example, in 1950, Nieman Marcus offered Black Angus steers for $2,230. The company only sold one animal to a gentleman in South Africa, and the company had to feed and care for it until it passed quarantine for shipment to its purchaser. In 1962, you could purchase a Chinese "junk-style" boat ($11,500), and in 1970, Noah's Ark ($588,247). For $10 the company would send you a sample of the wood that would be used to build your Ark.[2]

Some of the popular children's toys of the year that were first introduced in July 1969 and would appear in the upcoming *Sears Christmas Wish Book* included a board game called Ants in the Pants, a plastic version of tidily-winks, where players would attempt to launch large plastic ants into a pair of suspender-wearing trousers; Talking Barbie, who, after nine years of silence, finally spoke up with phrases like, "What should I wear to the prom?" and "I love being a fashion model"; and Battling Tops, where up to four players launched spinning tops into a concave arena where they would bounce off each other. The last top standing was the winner.

But in July 1969, the toy that was getting the most attention and would grace the pages of these catalogs in December was a plush doll called Snoopy Astronaut.

Created by cartoonist Charles Schulz, Snoopy was an adventurous beagle whose owner, Charlie Brown, was always on the losing end of life. Snoopy, on the other hand, had many careers over the years. He was a writer, although all of his stories began with, "It was a dark and stormy night." He was also a World War I flying ace, fighting a never-ending battle against the evil Red Baron.

Snoopy became inextricably linked to US space exploration with the advent of the Apollo program. During certain parts of the flights, Apollo astronauts would don specially designed communications caps. The caps were white with black earpieces that closely resembled the Charles Schulz beagle with his white body and black ears. The caps affectionately became known as Snoopy Caps.

Schulz's comic strip, *Peanuts*, and the strip's cast of characters built on the relationship with NASA with the launch of Apollo 10 in May 1969. The astronauts' command module was named *Charlie Brown*. The lunar module that astronauts Tom Stafford and Gene Cernan would test drive before the first lunar landing occurred two months later with Apollo 11 was nicknamed "Snoopy."

With this connection and the first lunar landing approaching, King Features (which syndicated the *Peanuts* comic strip) and Charles Schulz authorized the creation of a stuffed version of Snoopy, Snoopy Astronaut. Snoopy Astronaut came in his very own "flight safety" box and was dressed in a white Apollo space suit with a plastic helmet. Snoopy also carried a cooling unit much like the Apollo astronauts' own, and wore, that's right, a Snoopy Cap.

To this day, Snoopy is an important part of NASA and America's space effort. Each year, the agency awards the Silver Snoopy Award, a silver pin

An original 1969 Snoopy Astronaut doll dressed in his Apollo-era spacesuit and wearing the iconic Apollo "Snoopy Cap." *Sotheby's*

embossed with the famous beagle dressed in his Apollo spacesuit, to employees who contribute to flight safety and mission success.[3] Over the years, Snoopy Astronaut has made flights aboard the space shuttle and to the International Space Station. In August 2022, the Orion spacecraft lifted off from launch pad

39B at Cape Canaveral on a twenty-six-day mission to the moon as part of the Artemis project, the successor to Apollo that will land the next generation of men and women on the moon. The purpose of the unmanned Artemis 1 mission was to test all components of the Space Launch System before putting humans onboard and sending them to the moon. Onboard the capsule was a special passenger—an updated version of Snoopy Astronaut in an orange shuttle-era spacesuit. Snoopy had a specific job on this flight—he would act as a zero-gravity indicator. The toy was placed in the bottom of the capsule unattached to the vehicle. Cameras onboard Orion monitored the interior of the capsule, and when Snoopy was seen floating weightless, mission control knew that the capsule had reached zero-gravity.[4]

For your listening pleasure in the 1969 holiday season, there were two popular devices available—eight-track and cassette tapes. Both were extremely popular, and consumers could buy their favorite vinyl albums in either format, but both formats had their drawbacks.

The eight-track was invented in the late 1940s but didn't realize popularity until the late 1960s when the Ford Motor Company and later Chrysler and General Motors put eight-track players in their vehicles. Eight-tracks were bulky plastic cartridges that had a single loop of magnetic tape inside that contained recorded music. The tape had eight separate tracks recorded on it stacked one on top of the other, hence the name. A metal strip would tell the player when it was time to change to a new track. This is all well and good, except that many times, songs were not separated evenly on the tracks, so you may be listening to a song on track one and halfway through that song, it would pause, change to track two, and then continue playing the song. The format did, however, have better quality than its rival, the cassette tape.

Cassettes were much smaller in size, about the size of a deck of cards, which made them much more convenient to handle and store than an eight-track. Cassettes had a single magnetic tape inside, which allowed for an entire side of a standard vinyl album to be recorded onto it. No more having a song pause in the middle of it. Unlike the eight-track, however, when you came to the end of a side, you would have to manually flip the tape over.

The cassette tape was also well known for unspooling from the plastic container and getting jammed inside players. Thus, the trees along the highways of America were soon decorated with streamers of magnetic tape that frustrated owners had thrown out their car windows.

Despite that drawback, cassettes were much more popular than eight-tracks in July 1969 for their smaller size, convenience, and the ability to easily record your own music off the radio or record player, a phenomenon later known as the "mix tape."

For the rocket junkies out there, the flight of Apollo 11 and its predecessors opened the door for two new educational products that attempted to draw the youth of the world into careers in math and science. The first was called the Science Program, a project of Science Service and Nelson-Doubleday Publishing.

The Science Service was established in 1920 as a joint partnership between E.W. Scripps, the American Association for the Advancement of Science, and the National Research Council (NRC) as a news service designed to popularize science and to disseminate scientific knowledge. In the 1960s, the Science Service began its partnership with Doubleday, creating a series of sixty-four-page booklets designed to inspire future scientists. Topics included the weather, sound, crime detection, and space exploration.[5]

For only a dime, kids could begin their monthly subscription in the Science Service Program. *Society for Science*

The Man in Space series was first introduced in 1960 but was regularly updated through July 1969. For one thin dime, subscribers would receive the first sixty-four-page booklet, which was basically all text, save for a few random black-and-white sketches, a wall poster, and a series of full-color stamps that you could "paste" (lick the back like a postage stamp) in the book to bring the text to life. The series covered many aspects of spaceflight both manned and unmanned, digging into some fairly technical topics. The subscriber could examine the book for ten days, and if satisfied, a new booklet would be sent monthly complete with a maroon and silver faux leather binder to store the booklets in. Later editions also sent plastic models of capsules, rockets, even a model of the first American to walk in space, Ed White, tethered to a Gemini capsule.

But why simply read about flying rockets when you could build and actually fly one yourself? That was the goal of a small company in Penrose, Colorado, the Estes Model Rocket company: create a flying scale model rocket kit of the Apollo 11 Saturn V that could be built by kids and adults alike then taken out to their local ball field and actually launched into the sky. The result was one of the company's most popular model kits of all time.

Estes was the brainchild of Vern and Gleda Estes, who in 1958, created the company in the hopes of channeling hundreds of thousands of young people into productive roles in society and careers in space and science. Their effort included not only boys but girls as well, with their 1969 catalog showing a young girl hooking up the ignitors to one of their models on its portable launch pad.[6]

The company was known for the development of small, black powder–based rocket engines that were about the size of a roll of pennies and were extremely safe. The engines fit into a wide variety of the balsa-and-cardboard rocket kits that the company offered. Their designs ranged from detailed scale models of the early American Mercury Redstone rocket to futuristic science fiction based designs like their Mars Lander that resembled in many ways the Apollo lunar lander. Estes even sold still-image cameras that could be mounted to the top of a model rocket to take aerial photographs.

In their July 1969 catalog, the company introduced a highly detailed 1/100th scale model of the Apollo Saturn V rocket.[7] The finished kit stood 43.5 inches tall and used three Estes engines to launch it. To get the rocket off the ground, a special controller was needed that would connect a twelve-volt car battery and special metal wires (nichrome igniters) that were inserted into each engine. Push the button on the controller and the igniters would heat up and ignite the engines. Sometimes launching the model of the Saturn V was more challenging than building it. Many times one of the igniters would fail and only two of the three engines would ignite, sending the rocket off on an almost horizontal flight instead of vertical. But when all three ignited correctly, much like the real Saturn V, it was an impressive sight to see as it soared high into the sky.

The Estes were always very generous with their time and commitment to young people. Many of those who flew Estes model rockets in July 1969—both boys and girls—would go on to have careers in engineering and computer science or even become astronauts themselves.

"Gleda and I take great satisfaction in knowing that somehow we played a part [in the space program]," Vern Estes told *Launch Magazine* in October 2006. "Somewhere, somehow, each flight [of a NASA mission] carries with it the efforts of an Estes model rocketeer. Be it an astronaut, engineer, computer programmer, quality control inspector, program manager, or any of the many jobs and responsibilities involved, we know that somewhere in that chain is one or more of our 'youthful' customers of the 60s doing his or her job to make it happen."[8]

Leading up to the launch of Apollo 11, Vern and Gleda held a contest where one of their customers could join them for the actual launch at Cape Kennedy. The winner was fifteen-year-old Sven England of New Canaan, Connecticut, who joined the couple at the Cape for an experience of a lifetime.

For those looking for a snack to munch on while watching the first moon landing, people were buying up boxes of Fisher Moon Cheese by the thousands. "I don't know how we're going to handle the business," Carl Kaplanoff, director of Fisher's public relations department told United Press International. "We're getting so many orders our production people are in a quandary on how to step up production."[9]

The Fisher Cheese Company began in 1912 in the small Ohio town of Wapakoneta, the future hometown of Apollo 11 astronaut Neil Armstrong. Charles Fisher read about a little-known (at the time) breed of dairy cattle, the Dutch Holstein, and persuaded a local bank to give him a loan of $500 so that he could buy ten head.

The business slowly grew, starting with a single milk delivery route, then two, then it expanded to producing ice cream in 1920. In 1936, the Fishers decided it was time to expand once again, this time making ice cream and later, Moon Cheese.

Unlike the Moon Cheese that you see on grocery store shelves today that are dehydrated cheese puffs, Fisher Moon Cheese was actual American-made cheddar cheese.[10] The idea came to Charles Fisher after Apollo 8 first orbited the moon in 1968, and being the home of the first astronaut to walk on the moon, the idea only seemed natural.

By the time Apollo 11 was landing on the moon, word had spread about Moon Cheese, which came in a box that depicted an Apollo astronaut holding a round of the cheese with the words "unofficial sample" stenciled on it. Advertising for the cheese that appeared in newspapers and magazines depicted an Apollo command and service module and an astronaut floating in space with the caption, "Try Fisher Moon Cheese from the Hometown of Neil Armstrong the First Man on the Moon" proudly displayed on it.

Sales were brisk to say the least. "We're getting orders for Moon Cheese from all over the country," Kaplanoff continued. "We sent Florida over 50,000 pounds and they want more."[11]

At one point during the mission, Mission Control radios up to the Apollo 11 crew. "Eleven, Houston. A little information to you there. We've all taken a momentary respite from our work here to have . . . a bite of special moon cheese that is, I understand it has been sent to us directly from Wapakoneta, your own hometown."

Armstrong replied, "I think you'll enjoy that. They make a fine brand of cheese."[12]

There was no better advertising in the world—or out of it.

A FOOTNOTE TO HISTORY

July 29: French television expert Jean Felix-Charvet predicted that by 1985, ordinary TV receivers would be able to pick up satellite signals with the addition of a simple box. Felix-Charvet prophesied that those signals, in turn, could be used by adversarial countries to beam propaganda into the homes of smaller countries, breaking their will and causing mass confusion, chaos, and misinformation. That could never happen. Right? Felix-Charvet went on to urge international laws to prohibit such use of satellite technology.[13]

July 30: Alice Brock—who was the subject of the 1967 Arlo Guthrie album, *Alice's Restaurant*, in which Guthrie bemoaned the Vietnam War draft, was ticketed for disposing trash improperly on a roadside, and praising the restaurant where "you can get anything you want . . .

excepting Alice"—had closed her restaurant, but in 1969, she was moving on to two new careers. First, as an extra in the United Artist's movie that would be released later in the year based on the song. On July 30, it was announced that she was also publishing a new book—the *Alice's Restaurant Cookbook*.[14]

July 31: While rock bands were playing huge pop festivals around the world, including the largest, Woodstock, in Moscow, Russia, police reported that telephone booths were being vandalized by the city's youth, not necessarily for the money (although that was a nice side benefit), but rather for the parts—wires, mouthpieces, speakers, and the like— to change their acoustic guitars into electric guitars. Officials say the "advice" came from an article in an undisclosed magazine that provided tips on converting a standard six-string acoustic guitar to an electric by using parts from three telephones. The report described "tens of thousands" of Moscow phone booths being destroyed.[15]

EPILOGUE

What Rocket?

One would think that for the crew, the most grueling part of the flight of Apollo 11 was behind them once they were released from quarantine, but in fact, two months after being debriefed and taking some much-needed time off for a bit of rest and relaxation, Michael and Pat Collins, Buzz and Joan Aldrin, and Neil and Janet Armstrong would board a presidential plane along with an entourage from the US State Department and NASA officials to begin what was called the Giant Step World Tour.

On the morning of September 29, the Apollo 11 crew and their wives began a whirlwind world tour where they traveled to six continents, crossed the equator six times, and visited twenty-five cities and twenty-five countries in only thirty-eight days.[1] They met with many world leaders including Queen Elizabeth, the emperor of Japan, and the Shah of Iran, to name only a few, and they were deluged with gifts, including a nine-hundred-pound sculpture of Armstrong carved out of butter, presented to him by the Ohio Dairyman's Association.

At each stop, the crew was greeted with throngs of fans who wanted to get as close as possible to the returning heroes. They were treated as rock stars, with over one hundred million people turning out to greet them and an unimaginable amount of tickertape hurled from windows along the countless parades held in their honor.

The jubilation following the successful completion of President Kennedy's goal was short lived, however, and the world virtually turned away from the space program. The events of 1969—the Vietnam War, racism, poverty—were all-consuming issues, and more and more Americans lost interest in space, their gaze turning from the heavens back toward Earth and the seemingly

New York City welcomes Apollo 11 crewmen in a showering of ticker tape down Broadway and Park Avenue in a parade termed the largest in the city's history. Pictured in the lead car, from the right, are astronauts Neil A. Armstrong, commander; Michael Collins, command module pilot; and Edwin E. Aldrin Jr., lunar module pilot. *NASA*

insurmountable social and world problems the human race was facing. As Neil Armstrong lamented, "I had hoped the impact [of Apollo 11] would be more far-reaching. . . . We all seem to be sort of tied up with today's problems."[2]

Only four short months after the splashdown of Apollo 11, the next flight, Apollo 12 with astronauts Pete Conrad, Alan Bean, and Richard Gordon on board, rocketed into the sky from Cape Kennedy. The mission began ominously when the Saturn V was launched in the rain. As the towering rocket was clearing the launch tower, a bolt of lightning struck both the tower and spacecraft, momentarily knocking out the capsule's instrumentation. It could have been a monumental disaster, but the electronics on Apollo 12's capsule *Intrepid* recovered and the Saturn V continued to perform admirably.

Hardly anyone in the general public knew of the incident. By now, public opinion and the excitement of landing on the moon had vanished, ironically at a time when the US space program was hitting its zenith with the greatest achievement in mankind's history. The view of most Americans was that Apollo 11 was the culmination of the greater goal. The United States had beaten the Soviet Union to the moon. America had proven that they were technologically superior. Now, in the public eye at least, the Apollo program was over and had been completely shut down with the splashdown of Apollo 11's *Columbia* when in fact, six more missions were yet to be flown.

Public opinion of the Apollo project, and earlier crewed space-flight programs for that matter, had always been low. At the time of President Kennedy's commit-ment to land a man on the moon by the end of the decade, 60 per-cent of the American public were opposed to spending billions on

Sanitation workers cleaning the Apollo 11 parade route along Broadway, August 13, 1969. *Department of Sanitation Collection, NYC Municipal Archives*

a moon landing.[3] By the end of the Apollo 11 mission, polls had not changed much. While NASA and the astronauts received favorable ratings, 56 percent of Americans indicated that the United States should spend less on the space pro-gram, with only 10 percent saying that the budget should be increased.[4]

New York Times reporter Robert Reinhold was watching the further erosion of enthusiasm for the space program firsthand. In Nashville he watched as only a handful of people gazed at a small black-and-white television in a department store window that was tuned to a station covering the Apollo 12 launch. When he asked the manager of a local hotel newsstand whether she was interested in the Apollo 12 flight, she replied, "What rocket?"

Cedar Rapids, Iowa, attorney Tom Riley said, "You can't get as excited the second time you kiss the same girl as the first."[5]

And the opinion that the moon missions were diverting critical funding away from more earthly issues such as giving impoverished families a hand up contin-ued, growing even louder after Apollo 11 had landed.

"They've proved their point," Mrs. Elaine McClean from Tacoma, Wash-ington, was quoted as saying. "What the hell do they think they are doing now? That's our money, I tell you. I live in a poor neighborhood and see the needs."[6]

Even though the flight of Apollo 11 was one of the most momentous events in human history, one year after the mission, the public's short-term memory was kicking in. *New York Times* reporter Linda Charlton visited eight American cities and asked one simple question: What are the names of the Apollo 11 astronauts? While many of the respondents remembered the excitement of the first lunar landing, the bulk of them could not recall the astronauts' names. Out of ten people questioned in St. Louis, only four could remember Neil Armstrong's name. In New York, the number dropped to eight out of twenty-two who knew any of the astronauts' names.

When asked the name of the first man to walk on the moon, Paul Dickson of Pittsburgh, Pennsylvania replied, "It was Niles Something."[7]

Direct protests against the space program urging the government to do more to help the poor and hungry here on Earth began with the Poor People's Campaign March on Cape Kennedy the day before the launch of Apollo 11. Two weeks later, after splashdown, the movement was getting stronger and louder as it continued to grow in numbers despite the fact that funding for space exploration had already been drastically cut.

While the dollars spent on the Apollo program seemed astronomical to the average person, at its highest point of funding ($5 billion in 1965), the actual percentage of the US overall budget allocated to the space program was relatively small, at only 4.41 percent. The Vietnam War was inhaling most of American taxpayer's money. Even still, after 1965, the American space agency began to feel the pinch of budget cuts, with funding decreased to $4 billion in 1969 (2.31% of the US budget) and $3.7 billion in 1970 (1.92% of the budget).[8] By the 1990s, that figure had dropped to less than 1 percent of the federal budget.[9]

For Armstrong, Aldrin, and Collins, the first stop on their world tour was a state dinner held in their honor at the Century Plaza Hotel in Los Angeles. Over fourteen hundred foreign ambassadors, congressmen, entertainers, and Supreme Court justices arrived dressed in tuxedos and elegant gowns, all to hear from the first men on the moon.

Outside the hotel, there was a different gathering. On the street near the hotel's fountain, hundreds of young people wearing jeans, T-shirts, and sneakers had converged to be heard themselves, raising their voices in unison to ask questions of the government officials gathered inside. They asked why their friends had to die in an unpopular war. They chanted "Peace Now!" and held up signs reading "Bring the Troops Home Alive!"[10] They questioned why the United States could build a spectacular rocket but not keep the air we breathe and the water we drink clean. Every time one of the invited dinner guests stepped out of a limousine, the crowd would get louder and raise their fists in defiance.[11]

The crew of Apollo 11 was well aware of the social unrest being thrown at NASA, the space program, and Congress, having read about it in the newspapers delivered to them while in quarantine. A reporter covering the arrival of the astronauts in Los Angeles for the state dinner noticed a young African American boy kneeling in the crowd watching the astronauts entering the building. The reporter noted that Armstrong, seeing the boy, deliberately led the way as he, Aldrin, and Collins pushed through a sea of outstretched palms eagerly waiting to shake their hands so that they could shake the hand of the young man.[12]

As was the case with all returning Apollo astronauts, the city of New York would hold a ticker-tape parade in the astronaut's honor. The crew of Apollo 14—America's first astronaut, Alan Shepard along with Edgar Mitchell and Stuart Roosa—would receive the same red-carpet treatment in a parade that would be held on March 8, 1971.

The crew boarded a sixteen-car caravan outside the St. Regis Hotel and proceeded along the parade route flanked by a fleet of twenty-four siren-blaring, light-flashing police motorcycles. Thousands waved to the astronauts as the caravan passed, and tons of paper drifted from the monolithic city skyscrapers.

As the parade turned the corner at Broadway and West Third Street, it came to an abrupt halt. Twenty African American protestors who lived in the Broadway Central Hotel, a city housing unit for welfare recipients, had pushed baby strollers into the street. The protestors held signs that read, "White Astronauts Fly to the Moon While Black Children Die in Welfare Hotels" and chanted "Crumbs for Children, Millions for the Moon."[13] The blockade was quickly cleared by officers, but the protest continued at City Hall where Mayor John Lindsay presented the crew with gold medals.[14]

Even the Russians had become disenchanted with space, in particular the way in which their missions were kept in secrecy from the public unless the mission was a huge and historic success. "They treat us like children," one Moscow writer said. "Imagine, each day we learned for the first time that a different Soyuz ship was in space. Why could we not learn about it at once, in advance?"[15]

Russian scientists stressed to their leaders how important continued scientific space research was to the Soviet Union's economy, but the public was wary of the claims since their standard of living on Earth was bordering on poverty.

NASA began making plans for more economical Earth-bound programs that would still push the boundaries of science while fulfilling the promise administrator Thomas Paine made to Reverend Abernathy to bring the agency's innovative technology back to Earth. Plans were set in motion to build a reusable space vehicle, the Space Shuttle, that would eventually be used to build a permanent

Dr. Wernher von Braun, the first director of NASA's Marshall Space Flight Center, is carried aloft on the shoulders of city officials during the Apollo 11 celebration in downtown Huntsville on July 24, 1969. *NASA*

space station, the International Space Station (ISS). ISS would be an orbiting laboratory that would not only study the effects of extended spaceflights on humans, but also develop new drugs and technologies that could be used back on Earth to improve mankind's way of life.

Following Apollo 11's successful conclusion, the Rocket City, Huntsville, Alabama, went figuratively ballistic with excitement as they carried the man who made it all possible, Wernher von Braun, down the city's streets on their shoulders. But that excitement was quickly squelched with the realization that the dream was over.

The goal of landing men on the moon would not have been possible if it were not for the effort, skill, and intelligence of 118 German rocket scientists and engineers who were captured by American troops at the end of World War II. Their engineering skills created the V2 rocket, and while the rocket was used for terrifying and devastating results during the war, it would be the design that the Saturn V moon rocket would be based on and what made Huntsville the Rocket City.

In August 1970, those same engineers and thousands of contractors realized that even the history that they helped build with the Apollo program did not insulate them from what was coming. Of the 118 members of the German team, 80 were notified by NASA that their services would no longer be needed. Most of the men were in their late fifties and early sixties. The odds of them finding another job were small.

"Our jaws are dragging the floor," an unnamed wife of one of the engineers told the *New York Times*. "We didn't expect this blow, and we were not prepared for it."

"Somehow the old team felt perfectly safe in their jobs," Walter Weisman, the only non-technical member of the German team, said. "They didn't believe it could happen to them."[16]

The man behind the team, Wernher von Braun, was reassigned to Washington and a desk job with the space agency. His dream of landing men on the moon was only a jumping-off point to a bigger goal—landing humans on Mars, but his dreams were dashed with the cutbacks. Many in Huntsville believed that when von Braun died only eight years later, it was of heartbreak from the nation losing the vision.[17]

The mission of Huntsville's Marshall Spaceflight Center was diverted, focusing attention away from the moon to turning the third stage of a remaining Saturn V rocket into the first American space station, Skylab. With this decision, massive layoffs occurred, mainly with contractors to the Apollo program. Only two days after the launch of Apollo 11, a termination letter was sent to the Northrup Grumman Corporation to cease production of the lunar module, and North American Rockwell was told to suspend modifications to future Apollo command and service modules.[18]

By July 1970, several thousand NASA employees and contractors had been laid off, so many that astronauts on later Apollo missions began voicing their concerns that low morale was pervading the space agency's workers, which could jeopardize astronaut safety. America's first astronaut and Apollo 14 commander Alan Shepard feared that with a possible job loss looming over their heads, workers might say, "Maybe I won't be working here tomorrow so why should I worry about that little screw I just dropped down in a crack."[19]

Shepard's crewmate, Commander Edgar Mitchell, stressed that NASA has to "impress upon those that are worried about their jobs that we're still counting on them."[20]

While being laid off was a real possibility, the workers at all NASA facilities were dedicated to the mission. "I was amazed with the Cape [Kennedy] workers," lunar module test pilot Scott McCleod said. "They all knew they would be fired and that there was no reason to be there anymore. Everybody there kept right on working up to the last minute. There was no sabotage or people saying, 'I'll fix those guys!'"[21]

The town of Cocoa Beach, Florida, once a thriving mecca for space tourists, was also feeling the impact of the cutbacks. Storefronts were vacant with giant "For Rent" signs replacing sales banners in their windows. Hundreds of houses that once rang with the excitement of family life and the knowledge that these families were blazing new trails for America went silent and stood vacant.

"The Kennedy Space Center newsletter listed hundreds of houses for sale," contractor Shirley Wildermuth recalls. Shirley was one of the lucky ones who had kept her job. Her husband Wayne wasn't as lucky.

"I called one [of the listings] to see how much they wanted for it. The response was, 'You come over tonight and I'll sign it over to you because I'll be gone in the morning.' That was a common theme. You could pick up a house for nothing down and start making payments, but you were afraid to do so because you didn't know how long your job will last."[22]

As the final Apollo mission, Apollo 17, took off on December 7, 1972, in the only nighttime launch of the program, the hotels that were once booked to capacity up to fifty miles away from Cape Kennedy during the Apollo 11 launch were deserted. The ensuing years would look dismal for the towns around the Cape. The follow-up to Apollo, Skylab, didn't have the same impact as watching the first men head off to the moon, and the space shuttle would not launch for another nine years. When the first space shuttle finally did roll out to the launch pad twelve years after the launch of Apollo 11, the chair of the Brevard County Development Council, William Potter, put an optimistic spin on the moment. "Sales tax collections are up, bank deposits are up, and utility hookups are up. . . . In the long run, we're looking up."[23]

And, indeed, they were. The shuttle saw a resurgence of interest in the space program and the towns surrounding the Cape were beginning to blossom once again, not to the same fever pitch as the space program's early days, but still, they were making a comeback.

The space shuttle program gave NASA a rebirth as the agency learned how to partner with outside contractors and international partners to, as former NASA administrator and astronaut Daniel Goldin called it, "build faster, better, and cheaper," cutting costs while at the same time moving forward with cutting-edge technologies and bringing the science of space back to Earth.

NASA went on to keep administrator Thomas Paine's promise to Reverend Ralph Abernathy to use its incredible wealth of knowledge and technology to better life here on Earth. Since the flight of Apollo 11, the agency has converted many of its technologies to aid in everyday life, many of which we take for granted. The list seems endless: technology for food preservation, food poisoning prevention, improved osteoporosis treatments, water purification technologies, even technology that helps relieve symptoms of menopause.

On November 16, 2022, the United States once again began to reach for the moon at the renamed Cape Canaveral as the historic launch pad 39B, which had launched some of the Apollo astronauts to the moon from 1969 to 1972, came back to life with the Saturn V's successor, the Space Launch System (SLS), roaring to life just after midnight. SLS is part of the Artemis program, which intends to land the first woman and person of color on the moon.

Suddenly, the world was once again captured by the incredible power of the rocket and the stunning images its unmanned Orion capsule sent back to Earth of our fragile planet. In a poll taken by the Pew Research Center, 72 percent of Americans believed that the country should be the leader in space exploration even though, as was the case in the 1960s, only 42 percent believed the United States. should send people back to the moon. As before, the public believed that NASA's top priority should be focusing on solving the world's problems, including the ever-growing climate crisis.[24]

With the dawning of Artemis, a new space station—Gateway—has been planned to be built in lunar orbit, and with the continuation of research aboard the International Space Station and the ability to bring the technologies these efforts require back home to benefit all of us on Earth, the future of space travel looks brighter than it has in decades. And it was all built upon the shoulders of the events of July 1969.

ACKNOWLEDGMENTS

I could not have undertaken a project such as this without the help of family, some close friends, and a number of organizations. First and foremost, thank you once again to my wife, Maggie, for help putting the pieces of this book together and for putting up with me bouncing an endless number of questions and ideas off her. You don't know how much you've helped me in my writing journey. A special thanks to the NASA History Department for meticulously preserving all its records and making them easily accessible to the public. Special thanks to Jim Felder, Suzanne Hudson, and Joe Formichella for sharing their lives in 1969 for this book. And thank you to Sandra Friend and her husband, John Keatley. Sharing your connections with the workers that made Apollo happen was invaluable.

NOTES

INTRODUCTION

1. *The Moon above, the Earth Below*, producer Murial Pearson and Bernard Seabrooks, CBS Television Network, 1989, YouTube video, 1:35:46, https://youtu .be/EmT-MoiZ_I4.

2. *The Moon above, the Earth Below.*

CHAPTER 1

1. Joseph Lelyveld, "Only the Beaches Had the Look of the Fourth of July," *New York Times*, July 5, 1969, 1, 29.

2. Clark Whelton "Papo and the Hydrant: Lower East Side Summer," *Village Voice*, July 10, 1969, 1.

3. *The Moon above, the Earth Below*, producer Murial Pearson and Bernard Seabrooks, CBS Television Network, 1989, YouTube video, https://youtu.be/EmT -MoiZ_I4.

4. Nancy Moran, "Litchfield Salutes Its 250th Birthday," *New York Times*, July 5, 1969, 21.

5. *Mobile Press Register*, July 4, 1969, 1.

6. "Bill for 3-Day Holidays Becomes Law in Jersey," *New York Times*, July 4, 1969, 18.

7. "Deadly Weather: July 4, 1969 'Ohio Fireworks Derecho,'" Cleveland—WEWS TV/NewsNet 5, YouTube video, https://youtu.be/r6ITrsEMLq0.

8. Gary Rice, "45 Years Ago This Month . . . July 4th, 1969," *Lakewood Observer*, July 8, 2014, http://lakewoodobserver.com/read/2014/07/08/45-years-ago-this-month july-4th-1969.

9. "Nixon Greets Young Fans while Taking a Break in Leading July 4 Parade in Key Biscayne," *New York Times*, July 5, 1969, 1, 17.

10. "Flag on July 4: Thrill to Some, Threat to Others," *New York Times*, July 4, 1969, 23, 25.

11. Lelyveld, "Only the Beaches."

CHAPTER 2

1. U.S. Commission on Civil Rights, *Process of Change: The Story of School Desegregation in Syracuse, New York* (Washington, DC: Government Printing Office, 1968), 15. See https://www.google.com/books/edition/Process_of_Change/00oFAAAAMAAJ?hl=en&gbpv=1&bsq=riot.

2. U.S. Commission on Civil Rights, *Process of Change*, 15.

3. Joe Formichella, interview with the author, December 15, 2022.

4. Frank White, interview by Suzanne Wright, *Storycorps*, August 30, 2018. https://archive.storycorps.org/interviews/frank-whites-moon-landing-story/.

CHAPTER 3

1. Courtney G. Brooks, James M. Grimwood, and Lloyd S. Swenson Jr., *Chariots for Apollo: A History of Manned Lunar Spacecraft* (Washington, DC: US Government Printing Office, 1979), 329, https://www.hq.nasa.gov/office/pao/History/SP-4205/ch13-4.html#source2.

2. Anne M. Platoff, *Where No Flag Has Gone Before: Political and Technical Aspects of Placing a Flag on the Moon*, 1992, https://historycollection.jsc.nasa.gov/JSCHistoryPortal/history/flag/flag.htm.

3. Brooks et al., *Chariots for Apollo*, 330.

4. Brooks et al., 330.

5. Roger Simons, "Apollo 11 Left Disc of Poetic Messages and Humble Brags from Earth on Moon," *Orlando Sentinel*, July 8, 2019, https://www.orlandosentinel.com/space/apollo-11-anniversary/os-ne-apollo-11-messages-on-moon-20190708-czzqizb56jgcxca5wrqyuhlfdm-story.html.

6. Apollo 11 Goodwill Messages, NASA, July 13, 1969, https://www.history.nasa.gov/ap11-35ann/goodwill/Apollo_11_material.pdf.

7. Kiona N. Smith, "How Apollo 11 Raised the Flag on the Moon, and What It Means Today," *Forbes*, July 20, 2019, https://www.forbes.com/sites/kionasmith/2019/07/20/how-apollo-11-raised-the-flag-on-the-moon-and-what-it-means-today/?sh=6a3cb37e6f9e.

8. "Apollo Countdown," *Regina (Saskatchewan) Leader-Post*, July 2, 1969, 3.

9. Brooks et al., *Chariots for Apollo*, 327.

10. "'Eagle' to Land on Moon as 'Columbia' Waits," *Mobile Press Register*, July 6, 1969, 1.

11. "Apollo 11 Pre-launch Press Conference," NASA, Johnson Space Center, June 16, 1994, https://youtu.be/d_h8HCdV1oI.

12. Thomas J. Hamilton, "Soviet Space Debris Hits Japanese Ship Injuring 5," *New York Times*, July 5, 1969, 1, 27.

CHAPTER 4

1. Matthew Kirkman, "Prince Charles Fury: How 'Inappropriate' Prince of Wales Event Sparked Huge Backlash," *Daily Express*, March 4, 2019, https://www.express.co.uk/news/royal/1095418/prince-charles-news-inappropriate-investiture-parliament-protest-royal-news-spt.

2. United Press International, "Bomb Rocks Train Carrying Britain's Queen Elizabeth," *Lawton (Oklahoma) Constitution*, July 1, 1969, 1.

3. "Bomb Rocks Train Carrying Britain's Queen Elizabeth," *Lawton (Oklahoma) Constitution*, 1.

4. "John F. Kennedy and the Student Airlift," John F. Kennedy Presidential Library and Museum, https://www.jfklibrary.org/learn/about-jfk/jfk-in-history/john-f-kennedy-and-the-student-airlift.

5. "Ethnic Tensions Boil Over in Malaysia's 13 May 1969 Incident," Association for Diplomatic Studies, March 9, 2020, https://adst.org/2020/03/ethic-tensions-boil-over-in-malaysias-13-may-1969-incident/.

6. Kelly Jones, "A Time of Great Tension," *Berkeley Undergraduate Journal*, 76, https://escholarship.org/content/qt87n7c529/qt87n7c529_noSplash_33e60fb77dbf91886c97c487c82d9a66.pdf?t=nqarro.

7. Mitchell Lerner, "The Second Korean War," Wilson Center Digital Archive, December 14, 2009, https://digitalarchive.wilsoncenter.org/essays/second-korean-war.

CHAPTER 5

1. Len Buckwalter, "Buyers Guide to 1969 Color Sets," *Radio Electronics*, January 1969, 42, https://worldradiohistory.com/Archive-Radio-Electronics/60s/1969/Radio-Electronics-1969-01.pdf.

2. William Mitchell, "Hard Rock Rocked Hard," *National Catholic Reporter*, July 16, 1969.

3. John S. Wilson, "Unruly Rock Fans Upset Newport Jazz Festival," *New York Times*, July 7, 1969, 28.

4. Callum Crumlish, "The Beatles: John Lennon Drops Truth behind the Ballad of John and Yoko Tribute Song," *Daily Express*, September 5, 2020, https://www.express.co.uk/entertainment/music/1330789/the-beatles-john-lennon-the-ballad-of-john-and-yoko-paul-mccartney-song-recording.

5. "The Ballad of John and Yoko," *Beatles Bible*, February 28, 2022, https://www.beatlesbible.com/songs/the-ballad-of-john-and-yoko/.

6. Gil Kaufman, "Beatles Credit Feud Continues," *Rolling Stone Magazine*, December 16, 2003, https://www.rollingstone.com/music/music-news/beatles-credit-feud-continues-178487/.

7. "Brian Jones and Jim Morrison Die, Two Years Apart to the Day," *History Channel*, https://www.history.com/this-day-in-history/brian-jones-and-jim-morrison-die-two-years-apart-to-the-day#:~:text=Rolling%20Stones%20guitarist%20Brian%20Jones,drowning%20on%20July%203%2C%201969.

8. "Brian Jones: Sympathy for the Devil," *Rolling Stone Magazine*, August 9, 1969, https://www.rollingstone.com/music/music-news/brian-jones-sympathy-for-the-devil-182761/.

9. Al Delugach, "Kerkorian Agrees to Pay Summa $167 Million for the Sands and Desert Inn," *Los Angeles Times*, September 10, 1987, https://www.latimes.com/archives/la-xpm-1987-09-10-fi-6975-story.html.

10. G. David Wallace, "Living on Welfare Budget: Congressmen Try 18-Cents a Day," *Bedford (Indiana) Daily Times-Mail*, July 1, 1969, 12.

11. Leonard Kleinrock, "The First Message Transmission," *International Corporation for Assigned Names and Numbers (ICANN)*, October 29, 2019, https://www.icann.org/en/blogs/details/the-first-message-transmission-29-10-2019-en.

CHAPTER 6

1. "The POW Experience in the Vietnam War," https://www.vietnamwar50th.com/assets/1/7/VW50th_POW_Posters_7-10-19.pdf.

2. Rick Fredericksen, "How Troops in Vietnam Saw the First Moon Landing," Historynet, https://www.historynet.com/the-moonglow-of-apollo-11/.

3. John McDermott, "I Didn't Watch the Moon Landing," *Esquire*, July 20, 2019, https://www.esquire.com/news-politics/a28452621/moon-landing-stories/.

4. Fredericksen, "How Troops in Vietnam."

5. Kyle Peschler, "Vietnam Veterans Reflect: Apollo 11," https://www.njvvmf.org/home-depot-employees-pay-tribute-to-veterans-3/?utm_source=rss&utm_medium=rss&utm_campaign=home-depot-employees-pay-tribute-to-veterans-3.

CHAPTER 7

1. John F. Kennedy Presidential Papers, National Security Files, *Space Activities: US / USSR Cooperation 1961–1963*, https://www.jfklibrary.org/asset-viewer/archives/JFKNSF/308/JFKNSF-308-006.

2. John S. Wilson, "Podgorny Meets Borman, Voices Hope for Successful Moon Trip," *New York Times*, July 7, 1969, 20.

3. "US Exploration Legal, Moon Belongs to All," *Decatur (Illinois) Daily Review*, July 11, 1969, 1.

4. Courtney G. Brooks, James M. Grimwood, and Lloyd S. Swenson Jr., *Chariots for Apollo: A History of Manned Lunar Spacecraft* (Washington, DC: U.S. Government Printing Office, 1979), 329.

5. Brooks et al., 329.

6. Ivan D. Ertel and Roland W. Newkirk, *The Apollo Spacecraft Volume IV: January 21, 1966—July 13, 1974* (Washington, DC: US Government Printing Office, 1978), 306.

7. Brooks et al., *Chariots for Apollo*, 328.

8. Joseph E. Karth, "Letter to Dr. John S. Foster Jr.," National Reconnaissance Office, https://www.nro.gov/Portals/65/documents/foia/declass/mol/759.pdf, December 1, 2022.

9. US Congress, Proceedings of Congress and General Congressional Publications, 91st Congress, 1st sess., 1969, 20133, https://www.govinfo.gov/content/pkg/GPO -CRECB-1969-pt15/pdf/GPO-CRECB-1969-pt15-5-1.pdf.

10. Ertel and Newkirk, *The Apollo Spacecraft Volume IV*, 204.

11. "A New Ellington Score Marking Moon Landing," *New York Times*, July 12, 1969, 12.

12. Bernard Weinraub, "Tourists Crowd Cocoa Beach as Apollo Countdown Begins," *New York Times*, July 11, 1969, 1, 12.

13. Weinraub, 1, 12.

14. "Moon Launch Side Effects," *Park City (Utah) News*, July 13, 1969, 15.

CHAPTER 8

1. Michael P. Zboray, *Teenager's War: Vietnam 1969* (Zboray, 2019), 12.

2. Zboray, 15.

3. Zboray, 119.

4. "Vietnam War U.S. Military Fatal Casualty Statistics," National Archives, accessed July 7, 2022, https://www.archives.gov/research/military/vietnam-war/casu alty-statistics#:~:text=April%2029%2C%202008.-,The%20Vietnam%20Conflict%20 Extract%20Data%20File%20of%20the%20Defense%20Casualty,casualties%20of%20 the%20Vietnam%20War.

5. Joel Achenbach, "Did the News Media, Led by Walter Cronkite, Lose the War in Vietnam?," *Washington Post*, May 25, 2018, https://www.washingtonpost.com/ national/did-the-news-media-led-by-walter-cronkite-lose-the-war-in-vietnam/2018/05/ 25/a5b3e098-495e-11e8-827e-190efaf1f1ee_story.html.

6. *CBS Evening News*, "50 Years Ago: Walter Cronkite Calls for the U.S. to Get Out of Vietnam," YouTube video, November 1, 2022, 0:06 to 0:16, https://youtu.be/ Dn2RjahTi3M.

7. "Real America Preview: LBJ Announces He Won't Run 3/31/1968," CSPAN, You-Tube, November 1, 2022, news video, 0:35 to 0:50, https://youtu.be/CJeLoMCF6Jo.

8. "Nation Greets the First Troops Withdrawn by Nixon," *New York Times*, July 9, 1969, 1, 3.

9. "Nation Greets the First Troops Withdrawn by Nixon," 1, 3.

10. *The Moon above, the Earth Below,* producer Murial Pearson and Bernard Seabrooks, CBS Television Network, 1989, YouTube, https://youtu.be/EmT-MoiZ_I4.

11. *The Moon above, the Earth Below.*

12. "Resistance and Revolution: The Anti-Vietnam War Movement at the University of Michigan, 1965–1972," University of Michigan, accessed November 1, 2022, https://michiganintheworld.history.lsa.umich.edu/antivietnamwar/exhibits/show/exhibit/draft_protests/the-military-draft-during-the-#:~:text=On%20December%201%2C%201969%2C%20the,more%20frustrated%20with%20the%20system.

13. "A Call to Resist Illegitimate Authority," University of Massachusetts Boston, 1967, http://vietnamwar.lib.umb.edu/warHome/docs/1967CallToResistIllegit.html.

14. Associated Press, "Spock, Ferber Acquitted," *Troy (New York) Record*, July 12, 1969, 1.

15. DeNeen L. Brown, "'Shoot Them for What?' How Muhammed Ali Won His Greatest Fight," *Washington Post*, June 16, 2018, https://www.washingtonpost.com/news/retropolis/wp/2018/06/15/shoot-them-for-what-how-muhammad-ali-won-his-greatest-fight/.

16. Joe Pilati, "Five Women vs. the Draft Board," *Village Voice*, July 10, 1969, 24.

17. "Special to the Fifth Estate," Fifth Estate Collective, July 10–23, 1969, 1.

18. Pilati, "Five Women," 24.

19. "Concern for Vietnam on the Part of the American Friends Service Committee," American Friends Service Committee, 1969, https://www.afsc.org/sites/default/files/documents/1969%20AFSC%20Vietnam%20Timeline%2049-69.pdf.

20. "Concern for Vietnam."

21. Jonathan Black, "Gay Power Hits Back," *Village Voice*, July 31, 1969, 1.

22. Seth S. King, "Young Republicans Bid Nixon End War through a Victory," *New York Times*, 10.

CHAPTER 9

1. Federal Bureau of Investigation, *FBI Law Enforcement Bulletin*, Washington, DC: GPO, July 1969, https://www.fbi.gov/history/directors/j-edgar-hoover.

2. "85% Regard F.B.I. Favorably but Polls Find Some Slippage," *New York Times*, August 9, 1973, 24.

3. Federal Bureau of Investigation, "COINTELPRO," January 7, 2023, https://vault.fbi.gov/cointel-pro.

4. "African American Heritage: The Black Panther Party," National Archives, January 7, 2023, https://www.archives.gov/research/african-americans/black-power/black-panthers.

5. "How the Black Panther's Breakfast Program Both Inspired and Threatened the Government," History Channel, January 7, 2023, https://www.history.com/news/free-school-breakfast-black-panther-party.

6. "J. Edgar Hoover: Black Panthers Greatest Threat to U.S. Security," United Press International, July 16, 1969, https://www.upi.com/Archives/1969/07/16/J-Edgar-Hoover-Black-Panther-Greatest-Threat-to-US-Security/1571551977068/.

7. Federal Bureau of Investigation, "Subject: (COINTELPRO) Black Extremist 100-448006," July 14, 1969, https://vault.fbi.gov/cointel-pro/cointel-pro-black-extremists/cointelpro-black-extremists-part-14/view#document/p1.

8. Federal Bureau of Investigation.

9. Federal Bureau of Investigation.

10. Jacobi Williams, *From the Bullet to the Ballot: The Illinois Chapter of the Black Panthers* (Chapel Hill: University of North Carolina Press, 2013), 193.

11. Ayanna Archie, "The FBI Monitored Aretha Franklin's Role in the Civil Rights Movement for Years," NPR, September 12, 2022, https://www.npr.org/2022/09/12/1122319306/aretha-franklin-fbi-surveillance.

12. Jenn Dize and Afeni Evans, "Aretha Franklin Was Tracked by the FBI for 40 Years. Here's What's in Her File," *Rolling Stone*, October 2, 2022, https://www.rollingstone.com/music/music-features/aretha-franklin-fbi-file-surveillance-1234602217/.

13. Ayana Archie, "The FBI Monitored Aretha Franklin's Role in the Civil Rights Movement for Years," NPR, September 12, 2022, https://www.npr.org/2022/09/12/1122319306/aretha-franklin-fbi-surveillance.

14. Archie.

15. "Facts and Case Summary: *Miranda vs. Arizona*," United States Courts, January 7, 2023, https://www.uscourts.gov/educational-resources/educational-activities/facts-and-case-summary-miranda-v-arizona.

16. Department of Justice, "*Miranda vs. Arizona* Press Release," July 31, 1969, https://www.justice.gov/sites/default/files/ag/legacy/2011/08/23/07-31-1969b.pdf.

17. Roy Reed, "President Urges a National Drive on Narcotics Use," *New York Times*, July 15, 1969, 1, 18.

18. Legal Defense Fund, "Breonna Taylor, Amir Locke, and the Dangers of Warrant Executions," August 5, 2022, https://www.naacpldf.org/end-no-knock-warrants/#:~:text=Law%20enforcement's%20ability%20to%20execute,Reform%20and%20Criminal%20Procedure%20Act.

19. Andrew H. Malcolm, "Drug Law Change Sought by Percy," *New York Times*, November 4, 1973, 50.

20. Malcolm.

21. Malcolm.

22. Mino Marchese, "Examining the Risks of No-Knock Raids," American Legislative Exchange Council, April 14, 2022, https://alec.org/article/examining-the-risks-of-no-knock-raids/#:~:text=Currently%2C%2034%20states%20allow%20for,Virginia%20have%20banned%20them%20entirely.

CHAPTER 10

1. Bill DeMain, "The Sound and Vision of David Bowie," *Performing Songwriter*, September 2003.

2. Stephen Dowling, "How David Bowie Was Banned during the Moon Landing," *BBC Music*, July 16, 2019, https://www.bbc.com/culture/article/20190716-how-david -bowie-was-banned-during-the-moon-landing.

3. DeMain, "The Sound and Vision of David Bowie."

4. "Commander Chris Hadfield: David Bowie Really Liked How I Portrayed Space Oddity," *5 News Talk Live*, December 13, 2013, video, 6:06, https://youtu .be/_rTcIpWxy9I.

5. Becky Ferreira, "Chris Hadfield's Spirited Song in Space Was No Oddity," *New York Times*, November 2, 2020, https://www.nytimes.com/2020/11/02/science/chris -hadfield-space-oddity.html.

6. Jim Clash, "In the Year 2525, If Man Is Still Alive," *Forbes Magazine*, December 2, 2022, https://www.forbes.com/sites/jimclash/2020/04/03/in-the-year-2525-if-man-is -still-alive/?sh=671c350d9d2d.

7. Clash.

8. Peter Aaron, "Woodstock: A Legacy of Art and Community," *Chronogram*, August 1, 2019, https://www.chronogram.com/hudsonvalley/woodstock-a-legacy-of -art-and-community/Content?oid=8819944.

9. Aaron.

10. Bill Kovach, "Woodstock's a Stage, but Many Don't Care for the Show," *New York Times*, July 9, 1969, 45, 87.

11. Kovach.

12. "Napalm Lands on Freeway," *Lancaster (Pennsylvania) Intelligencer Journal*, July 8, 1969, 3.

13. "Understanding Your Language Rights," Office of the Commissioner of Official Languages, August 1, 2022, https://www.clo-ocol.gc.ca/en/language_rights/act.

14. "Newark Police Kill Rats," *Asbury Park Press*, July 9, 1969, 12, https://www .newspapers.com/image/144200771/.

CHAPTER 11

1. Jim Felder, interview with the author, October 4, 2022.

2. Felder.

3. Felder.

4. Lori Walters, "Shirley Wildermuth, Ground Support during Gemini, Apollo, Shuttle, Oral History 2002-11-01," American Space Museum, November 1, 2002, YouTube video, 58:08, https://youtu.be/oCTCIbj1EQg?list=PLPYxlf_GI2YIzAmky xm00LB-l5Ta2ysP4.

5. Walters.

6. Walters.

7. Mark Marquette, "Scott McCleod (Age 93) Veteran, Test Pilot, 'Lunar Module Test Astronaut,' Oral History 2019-04-25," American Space Museum, YouTube video, 37:52, April 25, 2019, https://youtu.be/uGgpwpXJ_DY?list=PLPYxlf_GI2YIzAmkyxm00LB-l5Ta2ysP4.

8. Marquette.

9. "Linda Vaden-Goad's Moon Landing Story," *Archive Storycorps*, audio, 4:12, September 21, 2018, https://archive.storycorps.org/interviews/linda-vaden-goads -moon-landing-story/.

10. "Linda Vaden-Goad's Moon Landing Story."

CHAPTER 12

1. "Firing Room 1 Learns Its Problems Are Few," *New York Times*, July 16, 1969, 20.

2. "Huntsville Keeps Head in Clouds," *Daily Advertiser*, July 13, 1969, 14.

3. Science and Technology Division, Library of Congress, *Astronautics and Aeronautics 1969* (Washington, DC: U.S. Government Printing Office, 1970), 206.

4. Courtney G. Brooks, James M. Grimwood, and Loyd S. Swenson Jr., *Chariots for Apollo: A History of Manned Lunar Spacecraft* (Washington, DC: US Government Printing Office, 1979), 334, https://www.hq.nasa.gov/office/pao/History/SP-4205/ch13-4.html#source2.

5. Brooks et al., 246.

6. Brooks et al., 334.

7. Exploring Stamps, "The Apollo 15 Scandal—S3E6," March 22, 2019, YouTube video, 15:31, https://youtu.be/yrkY_HB4ZS8.

8. "Firing Room 1 Learns Its Problems Are Few."

9. Roger Simmons, "Apollo 11 Countdown Started 50 Years Ago Today," *Orlando Sentinel*, July 10, 2019, https://www.orlandosentinel.com/space/apollo-11 -anniversary/os-ne-apollo-11-countdown-starts-20190710-tmbhprgdv5h6pevfloa3ei pege-story.html.

10. "Rocket Fuel in Her Blood: The Story of JoAnn Morgan," NASA History Department, July 12, 2019, https://www.nasa.gov/feature/the-story-of-joann-morgan.

11. "Rocket Fuel in Her Blood: The Story of JoAnn Morgan"

12. "Astrowives Prepared for This?," *Baltimore Evening Sun*, July 16, 1969, A2.

13. W. David Woods, Kenneth D. MacTaggart, and Frank O'Brien, "Apollo Flight Journal Day 1, Part 1: Launch," NASA History Department, October 3, 2022, https:// history.nasa.gov/afj/ap11fj/01launch.html#f1start.

14. Kristen Inbody, "NASA's Apollo 11 Moon Landing Is Montana's Story, Too," *Great Falls Tribune*, July 11, 2019, https://www.greatfallstribune.com/story/news/2019/07/10/nasas-apollo-11-moon-landing-montanas-story-too/1682898001/.

15. NASA Solar System Exploration, "Apollo 11," NASA, January 7, 2023, https:// solarsystem.nasa.gov/missions/luna-15/in-depth/.

16. Neil Hickey, "Live from the Moon," *T.V. Guide*, July 19, 1969, 7.

CHAPTER 13

1. Andrew Stanton, "Videos Show Iranians Celebrating U.S. Win in World Cup," *Newsweek*, November 29, 2022, https://www.newsweek.com/videos-show-iranians -celebrating-us-win-world-cup-1763331.

2. Michael McKnight, "Soccer. War. Nothing More," *Sports Illustrated*, June 3, 2019, https://www.si.com/soccer/2019/06/03/football-war-honduras-el-salvador.

3. McKnight.

4. Association for Diplomatic Studies and Training, "The 1969 'Soccer War' between Honduras and El Salvador," June 18, 2014, https://adst.org/2014/06/the-1969 -soccer-war/.

5. National Army Museum, "Ireland: Battle of the Boyne," January 7, 2023, https:// www.nam.ac.uk/explore/battle-boyne.

6. Jeff Wallenfeldt, "The Troubles: Northern Ireland History," *Britannica*, November 1, 2022, https://www.britannica.com/event/The-Troubles-Northern-Ireland -history.

7. John O'Brien, "Illuminations: The Spark That Lit the Fire—Burntollet Bridge," *iIrish*, January 30, 2019, https://iirish.us/2019/01/30/illuminations-the-spark-that-lit -the-fire-burntollet-bridge/.

CHAPTER 14

1. Sarah Pruitt, "Ted Kennedy's Chappaquiddick Incident: What Really Happened," *History Channel*, September 4, 2018, https://www.history.com/news/ted -kennedy-chappaquiddick-incident-what-really-happened-facts.

2. "RFK'S 'Boiler Room Girl' Was More Than Just Ted Kennedy's Rumored Blonde Lover," *Daily Mail*, April 3, 2018, https://www.dailymail.co.uk/news/fb -5575015/RFKS-BOILER-ROOM-GIRL-JUST-TEDDYS-RUMORED-BLOND -LOVER.html.

3. Leo Damore, *Chappaquiddick: Power, Privilege and the Ted Kennedy Cover-Up* (Washington, DC: Regnery History, 1988), 355.

4. Damore, 401.

5. Joseph Lelyveld, "Kennedy's Week of Tragedy," *New York Times*, July 27, 1969, 1.

6. Lelyveld, 1.

7. "Kennedys Attend Funeral," *The Sheboygan (Wisconsin) Press*, July 22, 1969, 1.

8. Damore, *Chappaquiddick*, 95.

9. United Press International, "Text of Kennedy's Appeal," *Atlanta Constitution*, July 26, 1969, 3.

10. United Press International, 3.

11. Associated Press, "Massachusetts in Uproar over Teddy's Statement," *Odessa (Texas) American*, July 26, 1969, 2.

12. Associated Press, "Dead Woman's Mother Hopes Kennedy Stays," *Odessa (Texas) American*, July 26, 1969, 2.

13. Associated Press, "Massachusetts in Uproar over Teddy's Statement," 1.

14. "Doctor Discusses Treatment of Nixon," *New York Times*, July 1, 1969, 25.

15. Robert D. McFadden, "Three Held in Plot on Nixon," *New York Times*, November 10, 1968, 1, 10.

16. "Police Here Check Arab Links of Accused in Nixon Plot," *New York Times*, November 12, 1968, 36.

17. Barnard L. Collier, "Doubt Is Raised on Nixon Plot," *New York Times*, November 15, 1968, 1, 32.

18. "Suspect in Plot to Assassinate Nixon Released in $25,000 Bail," *New York Times*, November 16, 1968, 19.

CHAPTER 15

1. Timothy Leary, "Tune in, Turn on, Drop Out," *The Library of Consciousness*, 1966, https://www.organism.earth/library/document/turn-on-tune-in-drop-out.

2. Roger Doughty, "Rock Rumbling into Rip Van Winkle Country," *Mobile Register*, July 9, 1969, 4C.

3. Doughty, 4C.

4. Richard F. Shepard, "Woodstock Festival Vows to Carry On," *New York Times*, July 18, 1969, 16.

5. Shepard.

6. "Woodstock Pop-Rock Fete Hits Snag," *New York Times*, July 17, 1969, 56.

7. Ray Cavanaugh, "Max Yasgur Rented Out His Farm for Woodstock. His Neighbors Sued Him," *Time*, August 14, 2019, https://time.com/5645555/woodstock-max-yasgur/.

8. "Youth: The Hippies," *Time*, July 7, 1969, https://content.time.com/time/magazine/0,9263,7601670707,00.html.

9. Molly Reid Clever, "Summer of Love 1969: When Love-Ins Came to New Orleans and New Orleans Went to Woodstock," *Historic New Orleans Collection*, September 1, 2022, https://www.hnoc.org/publications/first-draft/summer-love-1969-when-love-ins-came-new-orleans-and-new-orleans-went.

10. Clever.

11. John Stickney and John Olson, "The Youth Communes," *Life* magazine, July 18, 1969, 16A.

12. Stickney and Olson, 16A.

13. Stickney and Olson, 16A.

14. "'Starved' Local Youth Wins Soap Box Derby," *Lafayette Journal and Courier*, July 14, 1969, 1.

15. "32nd All-American Soap Box Derby Souvenir Program," All-American Soap Box Derby, Inc., August 23, 1969, https://klhess.com/soap-box-derby/SoapBoxDerby_SouvenirProgram_1969.pdf.

16. Editors of Encyclopedia Britannica, "Thor Heyerdahl," *Britannica*, January 7, 2023, https://www.britannica.com/biography/Thor-Heyerdahl.

17. Kon Tiki Museum, "On This Day, July 18, 1969," Facebook, July 18, 2018, https://www.facebook.com/KonTikiMuseum/posts/on-this-day-july-18th-1969-thor-heyerdahl-and-his-international-crew-on-the-reed/10155274103961230/.

18. "John Fairfax, First Man to Row Solo across the Atlantic, Dies Aged 74," *The Guardian*, February 20, 2012, https://www.theguardian.com/world/2012/feb/20/john-fairfax-first-row-atlantic.

19. Jeva Lange, "The Forgotten Adventurer Who Made History in Apollo 11's Shadow," *The Week*, July 19, 2019, https://theweek.com/articles/853544/forgotten-adventurer-who-made-history-apollo-11s-shadow#:~:text=Having%20stepped%20ashore%20at%20Hollywood,of%20the%20Apollo%2011%20astronauts.

CHAPTER 16

1. Interview of Jerry C. Elliott-High Eagle by David Zierler on October 2, 2020, Niels Bohr Library and Archives, American Institute of Physics, College Park, MD, www.aip.org/history-programs/niels-bohr-library/oral-histories/44901.

2. Elliott-High Eagle.

3. Elliott-High Eagle.

CHAPTER 17

1. Bob Granath, "Gemini's First Docking Turns into Wild Ride," NASA, March 3, 2016, https://www.nasa.gov/feature/geminis-first-docking-turns-to-wild-ride-in-orbit.

2. "50 Years Ago: Armstrong Survives Crash," NASA, May 7, 2018, https://www.nasa.gov/feature/50-years-ago-armstrong-survives-training-crash.

3. "Apollo Audio Gallery: We Have Liftoff," NASA, June 7, 2022, https://www.nasa.gov/mission_pages/apollo/apollo11_audio.html.

4. "Apollo Audio Gallery: We Have Liftoff."

5. Madalyn O'Hair vs. Thomas O. Paine, 312F. Supp. 434, United States District Court, W.D. Texas, Austin Division, December 1, 1969. https://law.justia.com/cases/federal/district-courts/FSupp/312/434/1468840/.

6. "Apollo 11: Celebrating Communion on the Moon," Guideposts, July 8, 2019, YouTube video, 5:26, https://www.youtube.com/watch?v=UQRvceotTMo.

7. Chris Carberry, *Alcohol in Space* (Jefferson, NC: McFarland & Company, 2019), 127.

8. "Webster Presbyterian Church History," Webster Presbyterian Church, n.d., https://www.websterpresby.org/content.cfm?id=329.

9. "Americans See Moon Landing as Peace Sign," *Reno Evening Gazette*, July 21, 1969, 6.

10. "Americans See Moon Landing as Peace Sign," 6.

11. "Letters: Readers Share Memories of Historic 1969 Moon Landing," *Orlando Sentinel*, July 20, 2019, https://www.orlandosentinel.com/space/apollo-11-anniversary/os-op-apollo-11-anniversary-moon-landing-memories-letters-20190720-wfayglvxvjgg pj2iqnqknfwmr4-story.html.

12. NASA Solar System Exploration, "Apollo 11," https://solarsystem.nasa.gov/missions/apollo-11/in-depth/.

13. Eric M. Jones, "Apollo 11 Surface Journal: EASEP Deployment and Closeout," NASA Headquarters, 1995, https://www.hq.nasa.gov/alsj/a11/a11.clsout.html.

14. David Kerley and Samantha Spitz, "50 Years Ago: The Pen That Saved Apollo 11," ABC News, July 12, 2019, https://abcnews.go.com/Politics/50-years-pen-saved-apollo-11/story?id=64228723.

15. Kerley and Spitz.

CHAPTER 18

1. "Weathering the Recovery of Apollo 11," NASA, August 16, 2019, https://www.nasa.gov/feature/weathering-the-recovery-of-apollo-11.

2. Jeremy Deaton, "'They Would Get Killed': The Weather Forecast That Saved Apollo 11," *Washington Post*, July 18, 2019, https://www.washingtonpost.com/weather/2019/07/18/weather-forecast-that-saved-apollo/.

3. "Weathering the Recovery of Apollo 11," NASA.

4. "Onboard the Qantas Flight Which Was Out of This World," Qantas Airlines, July 19, 2019, https://www.qantasnewsroom.com.au/roo-tales/onboard-the-qantas-flight-which-was-out-of-this-world/.

5. Gavin Mason, "Apollo 11—This Is Goddard—Including Qantas 596 Rare Film," QC Technologies, YouTube video, June 20, 2012, 14:56, https://youtu.be/Dp-nQVYehMk.

6. Sumit Singh, "The Time When Qantas Delayed a Flight to Watch the Return of Apollo 11," *Simple Flying*, April 27, 2020, https://simpleflying.com/qantas-apollo-11/.

7. "Remarks to Apollo 11 Astronauts Aboard the U.S.S. *Hornet* Following Completion of Their Lunar Mission," The American Presidency Project, July 24, 1969, https://www.presidency.ucsb.edu/documents/remarks-apollo-11-astronauts-aboard-the-uss-hornet-following-completion-their-lunar.

8. "Interview with Terry Slezak," by Rebecca Write, NASA Johnson Space Center Oral History Project, July 29, 2009, https://historycollection.jsc.nasa.gov/JSCHistoryPortal/history/oral_histories/SlezakTR/SlezakTR_7-29-09.htm.

9. "Woman Pricks Finger in Test of Moon Dust," *New York Times*, July 26, 1971, 54.

CHAPTER 19

1. Ben Evans, *Foothold in the Heavens: The Seventies, AmericaSpace* (Chichester, UK: Praxis Publishing, 2010), 167.

2. Kristen Inbody, "NASA's Apollo 11 Moon Landing Is Montana's Story, Too," *Great Falls Tribune*, July 10, 1969, https://www.greatfallstribune.com/story/news/2019/07/10/nasas-apollo-11-moon-landing-montanas-story-too/1682898001/.

3. "Lloyd Bott–Apollo 11," *HoneysuckleCreek.Net*, n.d., https://www.honeysuckle creek.net/supply/pioneers/lb_photos_a11.html.

4. Jeff Shesol, "Lyndon Johnson's Unsung Role Sending Americans to the Moon," *New Yorker*, July 20, 2019, https://www.newyorker.com/news/news-desk/lyndon -johnsons-unsung-role-in-sending-americans-to-the-moon.

5. Roger Simmons, "Apollo 11 Launch Included VIPs LBJ, Charles Lindbergh, Johnny Carson and Thousands More," *Orlando Sentinel*, July 2, 1969, https://www .orlandosentinel.com/space/apollo-11-anniversary/os-ne-apollo-11-vips-20190702 -alw64frfcfg47o7f2xap4hrjze-story.html.

6. Roger D. Launius, "Public Opinion Polls and Perceptions of US Human Space-flight," *Space Policy* 19, no 3 (2003): 163-75. Doi:10.1016/S0265-9646(03)00039-0, 2003, https://www.academia.edu/401035/_Evolving_Public_Perceptions_of_Human _Spaceflight_in_American_Culture_.

7. US Census Bureau, *24 Million Americans—Poverty in the United States: 1969*, accessed January 7, 2023, https://www.census.gov/library/publications/1970/demo/ p60-76.html.

8. Jane Van Niemen and Leonard C. Bruno, *NASA Historical Data Book: 1969–1978 Volume 4* (Washington, DC: US Government Printing Office, 1976), 6, https://his tory.nasa.gov/SP-4012/vol4/ch1.htm#:~:text=NASA's%20annual%20budget%2C%20 which%20had,considerable%20impact%20on%20the%20agency.

9. Richard Paul, "Washington Goes to the Moon Part 1," Public Radio Remix (PRX), May 13, 2019, audio, 52:01, https://beta.prx.org/stories/629.

10. Paul.

11. "To Save Spaceship Earth," *New York Times*, June 2, 1968, 181.

12. Science and Technology Division Library of Congress, *Astronautics and Aeronautics 1969* (Washington, DC: US Government Printing Office, 1970), 201.

13. Donald G. Ferguson, *JADARA Journal, Coordination: An Educator Views the Scene* (Fort Lauderdale: Nova Southeastern University Press, 2019), 40, https://nsu works.nova.edu/cgi/viewcontent.cgi?article=1493&context=jadara.

14. Drew Pearson, "Did Nixon's Doctor Get Permission to Report on Presidential Stability?," *Washington Post*, July 10, 1969, 11.

15. The Interviews, "Walter Cronkite on the 1969 Moon Landing," Television Academy Foundation, video, 28:51, https://interviews.televisionacademy.com/inter views/walter-cronkite.

16. The Interviews, "Walter Cronkite on the 1969 Moon Landing."

17. Andrew Chaikin, *Societal Impact of Spaceflight (NASA SP-2007-4801) Chapter 4: Live from the Moon—The Societal Impact of Apollo: Shifting Priorities* (Washington, DC: National Aeronautics and Space Administration Office of External Affairs, 2007), 4, https://history.nasa.gov/sp4801-chapter4.pdf.

18. Bryan Greene, "While NASA Was Landing on the Moon, Many African Americans Sought Economic Justice Instead," *Smithsonian Magazine*, July 11, 2019, https://www.smithsonianmag.com/history/nasa-landing-moon-many-african-americans-sought-economic-justice-instead-180972622/.

19. Hearings before the Subcommittee on Executive Reorganization of the Committee of Government Operations, Martin Luther King and Economic Justice 1966, United States Senate, Eighty-Ninth Congress, Second Session, December 15, 1966, Part 14, 2–3, https://college.cengage.com/history/ayers_primary_sources/king_justice_1966.htm.

20. "For One Instant, Apollo Unites Nation," *Chicago Defender*, July 21, 1969, 1, 3.

21. Thomas A. Johnson, "Blacks and Apollo: Most Couldn't Have Cared Less," *New York Times*, July 27, 1969, 143.

22. Neil M. Maher, *Apollo in the Age of Aquarius* (Cambridge, MA: Harvard University Press, 2017), 29.

23. Neil M. Maher, "Raised Fists and Lunar Rockets," *American Experience*, June 19, 2019, https://www.pbs.org/wgbh/americanexperience/features/chasing-moon-raised-fists-lunar-rockets/.

24. Science and Technology Division Library of Congress, *Astronautics and Aeronautics 1969*, 205.

25. "NASA Chief Briefs Abernathy after Protest at Cape," United Press International, July 16, 1969, https://www.upi.com/Archives/1969/07/16/NASA-chief-briefs-Abernathy-after-protest-at-Cape/7371558396299/.

26. Thomas O. Paine, "Memorandum for Record" (official memorandum, Washington, DC: National Aeronautics and Space Administration, 1969).

27. Paine.

28. "NASA Chief Briefs Abernathy."

29. Paine, "Memorandum for Record."

30. Marcus Baram, *Gil Scott-Heron: Pieces of a Man* (New York: St. Martin's Press, 2014), 75.

CHAPTER 20

1. *Report of the National Advisory Commission on Civil Disorders, Summary of Report* (New York: Bantam Books, 1968), 1–29, http://www.eisenhowerfoundation.org/docs/kerner.pdf.

2. "Rally to Focus on Problems of Poor," *Indianapolis Star*, July 12, 1969, 8.

3. "Rally to Focus on Problems of Poor," 8.

4. "Marchers Climb Mansion Walls, Calm Restored," *Indianapolis Star*, July 14, 1969, 2.

5. George Lindberg, "Mansion Marches Futile: Whitcomb," *Indianapolis Star*, July 29, 1969, 19.

6. Lindberg, 19.

7. Martin Waldron, "Six Negroes Win Alabama Offices," *New York Times*, July 30, 1969, 1, 24.

8. Waldron.

9. "Stonewall through the Years," *American Experience*, n.d., https://www.pbs .org/wgbh/americanexperience/features/stonewall-inn-through-years/.

10. "Why Did the Mafia Own the Bar?," *American Experience*, n.d., https:// www.pbs.org/wgbh/americanexperience/features/stonewall-why-did-mafia-own -bar/#:~:text=At%20the%20time%20of%20the,by%20the%20New%20York%20Mafia .&text=Organized%20crime%20families%20owned%20the,lasted%20throughout%20 the%20late%201960s.

11. "The Birthplace," The Stonewall Inn, April 4, 2017, https://thestonewallinnnyc .com/the-stonewall-inn-story/2017/4/4/ntmsg5ni7iixxdjimmg16hz6dvsi4v.

12. Jerry Lisker, "Homo Nest Raided, Queen Bees Are Stinging Mad," *New York Sunday News*, July 6, 1969, 1.

13. "The Birthplace."

14. Jonathan Black, "Gay Power Hits Back," *Village Voice*, July 31, 1969, 28.

15. Black, 28.

16. Black, 28.

CHAPTER 21

1. "Sports," *Current History Magazine*, June 30, 1939, 53.

2. "Aretha Franklin Fined $50," *New York Times*, July 26, 1969, 40.

3. Mark Beaumont, "The Who's '*Tommy*': An In-Depth Look at Their Ground-breaking Rock Opera Album," *New Musical Express*, September 22, 2016, https://www .nme.com/blogs/nme-blogs/the-whos-tommy-anatomy-1554947.

4. Michael Davis, *Street Gang: The Complete History of Sesame Street* (London: Penguin Books, 2009), 161.

5. Davis, 159.

6. John D. Morris, "Tobacco Industry Pledges Broadcast Ban in 1970," *New York Times*, July 23, 1969, 1, 19.

CHAPTER 22

1. Letter of Samuel C. Phillips to George M. Low, "Control and Disposition of Apollo 11 Hardware," NASA Headquarters, July 28, 1969.

2. "Moon Rocks Story," *New York Times*, July 30, 1969, 38.

3. Lindsey Hamish, "ALSEP: Apollo Lunar Surface Experiments Package, 19 November 1969–30 September 1977," NASA, https://www.hq.nasa.gov/alsj/HamishALSEP.html, July 10, 2010.

4. Jack Gould, "TV: Pictures Raise Questions about Moon's Color," *New York Times*, July 30, 1969, 79.

5. "Apollo 11 Anniversary: Armstrong's Code," *Washington Post*, https://www.washingtonpost.com/wp-srv/national/longterm/space/armstrong4.htm.

6. "How Moon Landing Conspiracy Theories Began and Why They Persist Today," University of Manchester, July 12, 2019, https://www.manchester.ac.uk/discover/news/moon-landing-conspiracy-theories/.

7. Olivia McKelvey, "Conspiracy Theorist Punched by Buzz Aldrin Still Insists Moon Landing Was Fake," *Florida Today*, July 19, 2019, https://www.floridatoday.com/story/tech/science/space/2019/07/19/lunar-landing-denier-we-never-went-moon/1702676001/.

8. Science and Technology Division Library of Congress, Astronautics and Aeronautics 1969 (Washington, DC: US Government Printing Office, 1970), 206, https://history.nasa.gov/AAchronologies/1969.pdf.

9. Thomas Paine, "Foreign Relations, 1969–1976, Volume E-1, Documents on Global Issues, 1969–1972," US Department of State Archives, August 22, 1969, https://2001-2009.state.gov/r/pa/ho/frus/nixon/e1/46093.htm.

10. NASA, Astronautics and Aeronautics 1969 (Washington, DC: Science and Technology Division Library of Congress).

11. Johannes Kemppanen, "Apollo Flight Journal: More Than SCE to AUX—Apollo 12 Lightning Strike Incident," NASA History Department, February 24, 2021, https://history.nasa.gov/afj/ap12fj/a12-lightningstrike.html.

12. Kemppanen.

13. Courtney G. Brooks, Ivan D. Ertel, and Roland W. Newkirk, "SP-4011 Skylab: A Chronology, Part II-Apollo Application Program (Continued)," NASA History Department, n.d., https://history.nasa.gov/SP-4011/part2c.htm.

14. Editors, "What Price Moon Dust?," *The Nation*, July 28, 1969, https://www.thenation.com/article/archive/what-price-moondust/.

CHAPTER 23

1. Lauren-Brooke Eisen, "America's Faulty Perception of Crime Rates," Brennan Center for Justice, March 16, 2015, https://www.brennancenter.org/our-work/analysis-opinion/americas-faulty-perception-crime-rates.

2. "United States Population and Number of Crimes 1960–2019," The Disaster Center, n.d., https://www.disastercenter.com/crime/uscrime.htm.

3. Casey Sherman, *Helltown: The Untold Story of a Serial Killer on Cape Cod* (Naperville, IL: Sourcebooks, 2022), 93.

4. Sherman, 107.

5. Sherman, 183.

6. Neil Patmore, "The Depraved Crimes of Tony Costa, the 'Cape Cod Vampire' Who Terrorized Massachusetts in the 1960s," All That's Interesting, October 4, 2022, https://allthatsinteresting.com/tony-costa.

7. Kurt Vonnegut Jr., "A Novelist's Tale of Four Horrible Cape Cod Murders: 'There's a Maniac Loose Out There,'" *Life*, July 25, 1969, 53.

8. Vonnegut, 53.

9. Michael F. Cole, *The Zodiac Revisited: Volume 1—The Facts of the Case* (Northbrook, IL: Two Prime Publishing 2020), 510.

10. Cole, 521.

11. Andrew Blankstein and Wilson Wong, "'The Case Remains Open': FBI Rebuts Claim Zodiac Killer Case Is Solved," NBC News, October 7, 2021, https://www.nbc news.com/news/us-news/case-remains-open-fbi-refutes-claim-zodiac-killer-case-solved -n1281002.

12. Jeff Gunn, *Manson: The Life and Times of Charles Manson* (New York: Simon and Schuster, 2013), 79.

13. Vincent Bugliosi, *Helter Skelter: The True Story of the Manson Murders* (New York: W.W. Norton, 1974), 225.

14. Bugliosi, 225.

15. Leslie Kennedy, "How Spahn Ranch Became a Headquarters for the Manson Family Cult," History Channel, August 8, 2019, https://www.history.com/news/spahn -ranch-manson-family.

16. Adam Carlson and Christine Pelisek, "Relative of First Manson 'Family' Victim Speaks Out after Cult Leader's Death: 'About Time,'" *People*, November 21, 2017, https://people.com/crime/gary-hinman-family-charles-manson-death/.

CHAPTER 24

1. Brian Alexander, "Free Love: Was There a Price to Pay?," NBC News, June 22, 2007, https://www.nbcnews.com/id/wbna19053382.

2. "Modern Living: Sex as a Spectator Sport," *Time*, July 11, 1969, http://content .time.com/subscriber/article/0,33009,901005,00.html.

3. Russell Baker, "Don't Ask Me, I Only Live Here," *Life*, July 25, 1969, 35.

4. "1969 in Film," The Culture Wiki, n.d., https://culture.fandom.com/wiki/ 1969_in_film.

5. Amir Khafagy, "Goodbye, Show World: The Last Days of Times Square's Peep Shows," *Curbed New York*, April 25, 2019, https://ny.curbed.com/2019/4/25/ 18306448/times-square-new-york-peep-show-history.

6. "From Dazzling to Dirty and Back Again: A Brief History of Times Square," Times Square: The Official Website, n.d., https://www.timessquarenyc.org/history-of -times-square.

7. Alexis Soloski, "'*Oh! Calcutta!*' at 50: Still Naked After All These Years," *New York Times*, June 17, 2019, https://www.nytimes.com/2019/06/17/theater/oh-calcutta -at-50.html.

8. Clive Barnes, "Theater: '*Oh! Calcutta!*' a Most Innocent Dirty Show," *New York Times*, June 18, 1969, 33.

9. A Book Named "John Cleland's Memoirs of a Woman of Pleasure" et al., Appellants vs. Attorney General of the Commonwealth of Massachusetts, No. 368 (U.S. Supreme Court, 1966), https://www.law.cornell.edu/supremecourt/text/383/413.

10. Jacqueline Susann, *The Love Machine* (New York: Simon and Schuster, 1969).

11. Nora Ephron, "The Love Machine," *New York Times*, May 11, 1969, 101, 110.

12. Ephron.

CHAPTER 25

1. "The Nieman Marcus Christmas Book 1969," ComputerHistory.Org, 1969, 84, http://archive.computerhistory.org/resources/access/text/2011/01/102685475-03 -01-acc.pdf.

2. Julee Wilson, "The Most Jaw Dropping Fantasy Gifts in the History of Nieman Marcus," *Huffington Post*, November 25, 2015, https://www.huffpost.com/entry/jaw -dropping-neiman-marcus-fantasy-gifts_n_56339abee4b0c66bae5c2e7b.

3. Spaceflight Awareness, "Silver Snoopy Award," NASA, n.d., https://www.nasa .gov/directorates/heo/sfa/aac/silver-snoopy-award.

4. Artemis 1, "Snoopy to Fly on NASA's Artemis 1 Moon Mission," NASA, November 12, 2021, https://www.nasa.gov/feature/snoopy-to-fly-on-nasas-artemis -i-moon-mission.

5. "1960s: The Science Program," Society for Science, n.d., https://centennial.soci etyforscience.org/entry/1960s-the-science-program/.

6. "Space Age Legends: Vern and Gleda Estes Reflect Back on the Hey Day of Model Rocketry," *Launch Magazine*, September/October 2006, 25.

7. "Estes Model Rocket Catalog," Estes Model Rocket Company, 1969, 23.

8. "Space Age Legends: Vern and Gleda Estes," 61.

9. United Press International, "Wapakoneta Moon Cheese Operation Just Booming," UPI Archives, July 17, 1969, https://www.upi.com/Archives/1969/07/17/Wapa koneta-moon-cheese-operation-just-booming/4251557961411/.

10. Dave Baeumel, "Moonwalk Memories: Oakwood Man Lived Just a 'Few Farms' from Neil Armstrong," *Gainesville Times*, September 23, 2011, https://www.gainesville times.com/news/moonwalk-memories-oakwood-man-lived-just-a-few-farms-from-neil -armstrong/.

11. "Maker of 'Moon Cheese' Reports Heavy Demand," *New York Times*, July 18, 1969, 13.

12. Apollo Flight Journal, Day 3, Part 3: Flight Plan Updates, https://history.nasa .gov/afj/ap11fj/10day3-flight-plan-update.html.

13. "Satellite Seen Useful for TV Propaganda," *New York Times*, July 30, 1969, 18.

14. Judy Klemesrud, "A Cookbook by That Restaurant's Alice," *New York Times*, July 30, 1969, 44.

15. James F. Clarity, "Moscow Phone Booths Are Prey of Guitarists and Coin Thieves," *New York Times*, July 31, 1969, 7.

EPILOGUE

1. "Apollo 11 Anniversary: Armstrong's Code," *Washington Post*, https://www.washingtonpost.com/wp-srv/national/longterm/space/armstrong4.htm.

2. "Apollo 11 Anniversary: Armstrong's Code."

3. Neil M. Maher, *Apollo in the Age of Aquarius* (Cambridge, MA: Harvard University Press), 29.

4. Maher, 289.

5. Robert Reinhold, "Enthusiasm for Lunar Exploration Is Found to be on Wane around Country," *New York Times*, November 15, 1969, 23.

6. Reinhold, 23.

7. Linda Charlton, "Check Finds Many Forget Apollo 11," *New York Times*, July 19, 1970, 54.

8. Jane Van Niemen and Leonard C. Bruno, *NASA Historical Data Book: 1969–1978 Volume IV* (Washington, DC: US Government Printing Office, 1976), 6, https://history.nasa.gov/SP-4012/vol4/ch1.htm#:~:text=NASA's%20annual%20budget%2C%20which%20had,considerable%20impact%20on%20the%20agency.

9. "NASA Budgets: US Spending on Space Travel Since 1958 Updated," *The Guardian*, n.d., https://www.theguardian.com/news/datablog/2010/feb/01/nasa-budgets-us-spending-space-travel.

10. Steven V. Roberts, "Astronauts Find Mixed Reactions," *New York Times*, August 15, 1969, 14.

11. Roberts, 14.

12. Robert Stone and Alan Andres, *Chasing the Moon: The People, the Politics, and the Promise That Launched America into the Space Age* (New York: Ballantine Books 2019), 282.

13. Neil M. Maher, "Raised Fists and Lunar Rockets," *American Experience*, June 19, 2019, https://www.pbs.org/wgbh/americanexperience/features/chasing-moon-raised-fists-lunar-rockets/.

14. Paul L. Montgomery, "Protests Interrupt City Welcome for Astronauts," *New York Times*, March 9, 1971, 1, 34.

15. Bernard Gwertzman, "Russians Seem Disenchanted with Space Program," *New York Times*, November 15, 1969, 24.

16. John Noble Wilford, "NASA Layoffs Hit Von Braun Team," *New York Times*, September 3, 1970, 1, 22.

17. Jim Felder interview with the author, October 4, 2022.

18. Courtney G. Brooks, Ivan D. Ertel, and Roland W. Newkirk, *SP-4010: Skylab-A Chronology: Part II: Apollo Application Program* (Washington, DC: US Government Printing Office, 1976), 177, https://history.nasa.gov/SP-4011/part2c.htm.

19. "Apollo Crew Upset on Worker Morale," *New York Times*, July 12, 1970, 67.

20. "Apollo Crew Upset on Worker Morale," 67.

21. Mark Marquette, "Scott McCleod (Age 93) Veteran, Test Pilot, 'Lunar Module Test Astronaut,' Oral History 2019-04-25," American Space Museum, YouTube video, April 25, 2019, https://youtu.be/uGgpwpXJ_DY?list=PLPYxlf_GI2YIzAmkyxm00LB-l5Ta2ysP4.

22. Marquette.

23. Boyce Rensbergers, "Cape Hoping for the Best in Wake of Apollo's Boom," *New York Times*, December 8, 1979, 30.

24. Courtney Johnson, "How Americans See the Future of Space Exploration, 50 Years after the First Moon Landing," Pew Research Center, July 17, 2019, https://www.pewresearch.org/fact-tank/2019/07/17/how-americans-see-the-future-of-space-exploration-50-years-after-the-first-moon-landing/.

INDEX

ABOUT THE AUTHOR

Joe Cuhaj grew up in New Jersey as a space fanatic, skipping school to watch every launch and recovery from the late Mercury missions to the final Skylab mission, but he wasn't a rocket geek. He loved the inside, personal stories that made manned exploration of space special. On any given Sunday, you would find him flying model rockets in fields with his friends in northern New Jersey. Cuhaj is a navy veteran and former radio broadcaster turned author and freelance writer. During his radio career, Joe applied to take part in NASA's Journalist in Space program, but he never heard back.

He is the author of twelve books, including *Hiking Alabama, 5th Edition* (2022), *A History Lover's Guide to Mobile and the Alabama Gulf Coast* (2023), and *Space Oddities: Forgotten Stories of Mankind's Exploration of Space* (2022).